APPLIED METHODOLOGIES IN POLYMER RESEARCH AND TECHNOLOGY

APPLIED METHODOLOGIES IN POLYMER RESEARCH AND TECHNOLOGY

Edited by
Abbas Hamrang, PhD, and Devrim Balköse, PhD

Gennady E. Zaikov, DSc, and A. K. Haghi, PhD
Reviewers and Advisory Board Members

Apple Academic Press
TORONTO   NEW JERSEY

Apple Academic Press Inc. | Apple Academic Press Inc.
3333 Mistwell Crescent | 9 Spinnaker Way
Oakville, ON L6L 0A2 | Waretown, NJ 08758
Canada | USA

©2015 by Apple Academic Press, Inc.

First issued in paperback 2021

Exclusive worldwide distribution by CRC Press, a member of Taylor & Francis Group
No claim to original U.S. Government works

ISBN 13: 978-1-77463-356-4 (pbk)
ISBN 13: 978-1-77188-040-4 (hbk)

Library of Congress Control Number: 2014950955

Library and Archives Canada Cataloguing in Publication

Applied methodologies in polymer research and technology/edited by Abbas Hamrang, PhD, and Devrim Balköse, PhD; Gennady E. Zaikov, DSc, and A.K. Haghi, PhD Reviewers and Advisory Board Members.

Includes bibliographical references and index.
ISBN 978-1-77188-040-4 (bound)
1. Polymers. I. Hamrang, Abbas, editor II. Balköse, Devrim, author, editor

TA455.P58A66 2014   620.1'92   C2014-906464-0

Apple Academic Press also publishes its books in a variety of electronic formats. Some content that appears in print may not be available in electronic format. For information about Apple Academic Press products, visit our website at **www.appleacademicpress.com** and the CRC Press website at **www.crcpress.com**

ABOUT THE EDITORS

Abbas Hamrang, PhD

Abbas Hamrang, PhD, is a professor of polymer science and technology. He is currently a senior polymer consultant and editor and member of the academic boards of various international journals. His research interests include degradation studies of historical objects and archival materials, cellulose-based plastics, thermogravemetric analysis, and accelerated ageing process and stabilization of polymers by chemical and non-chemical methods. His previous involvement in academic and industry sectors at the international level includes deputy vice-chancellor of research and development, senior lecturer, manufacturing consultant, and science and technology advisor.

Devrim Balköse, PhD

Devrim Balköse, PhD, graduated from the Middle East Technical University in Ankara, Turkey, with a degree in chemical engineering. She received her MS and PhD degrees from Ege University, Izmir, Turkey, in 1974 and 1977 respectively. She became associate professor in macromolecular chemistry in 1983 and professor in process and reactor engineering in 1990. She worked as research assistant, assistant professor, associate professor, and professor between 1970–2000 at Ege University. She was the Head of Chemical Engineering Department at Izmir Institute of Technology, Izmir, Turkey, between 2000 and 2009. She is now a faculty member in the same department. Her research interests are in polymer reaction engineering, polymer foams and films, adsorbent development, and moisture sorption. Her research projects are on nanosized zinc borate production, ZnO polymer composites, zinc borate lubricants, antistatic additives, and metal soaps.

REVIEWERS AND ADVISORY BOARD MEMBERS

Gennady E. Zaikov, DSc

Gennady E. Zaikov, DSc, is Head of the Polymer Division at the N. M. Emanuel Institute of Biochemical Physics, Russian Academy of Sciences, Moscow, Russia, and Professor at Moscow State Academy of Fine Chemical Technology, Russia, as well as Professor at Kazan National Research Technological University, Kazan, Russia. He is also a prolific author, researcher, and lecturer. He has received several awards for his work, including the Russian Federation Scholarship for Outstanding Scientists. He has been a member of many professional organizations and on the editorial boards of many international science journals.

A. K. Haghi, PhD

A. K. Haghi, PhD, holds a BSc in urban and environmental engineering from University of North Carolina (USA); a MSc in mechanical engineering from North Carolina A&T State University (USA); a DEA in applied mechanics, acoustics and materials from Université de Technologie de Compiègne (France); and a PhD in engineering sciences from Université de Franche-Comté (France). He is the author and editor of 65 books as well as 1000 published papers in various journals and conference proceedings. Dr. Haghi has received several grants, consulted for a number of major corporations, and is a frequent speaker to national and international audiences. Since 1983, he served as a professor at several universities. He is currently Editor-in-Chief of the *International Journal of Chemoinformatics and Chemical Engineering* and *Polymers Research Journal* and on the editorial boards of many international journals. He is a member of the Canadian Research and Development Center of Sciences and Cultures (CRDCSC), Montreal, Quebec, Canada.

CONTENTS

LIST OF CONTRIBUTORS

Esen Arkış
Izmir Institute of Technology Department of Chemical Engineering, Gülbahce Urla 35430, Izmir Turkey, Email: esenarkis@iyte.edu.tr

Devrim Balköse
Izmir Institute of Technology Department of Chemical Engineering, Gülbahce Urla 35430 Izmir Turkey

Farshid Basiri
Department of Chemical Engineering, South Tehran Branch, Islamic Azad University, Tehran, Iran

Liliya Bazylyak
Chemistry of Oxidizing Processes Division; Physical Chemistry of Combustible Minerals Department, Institute of Physical–Organic Chemistry & Coal Chemistry named after L. M. Lytvynenko, National Academy of Science of Ukraine 79053, Ukraine, Email: bazyljak.L.I@nas.gov.ua

Alireza Eslami
Department of Chemical Engineering, South Tehran Branch, Islamic Azad University, Tehran, Iran

M. Hasanzadeh
Department of Textile Engineering, University of Guilan, Rasht, Iran

Mahdi Hasanzadeh
Department of Textile Engineering, University of Guilan, Rasht, Iran, Email: hasanzadeh_mahdi@yahoo.com

Aleksei A. Iordanskii
Semenov Institute of Chemical Physics, Russian Academy of Sciences, Moscow, Russia

Svetlana G. Karpova
Emmanuel Institute of Biochemical Physics, Russian Academy of Sciences, Moscow, Russia

Azamat A. Khashirov
Kabardino-Balkarian State University, Nalchik 360004, Russia, Russian Federation, Email: new_kompozit@mail.ru

Svetlana Yu. Khashirova
Kabardino-Balkarian State University, Nalchik 360004, Russia, Russian Federation, Email: new_kompozit@mail.ru

Sergei M. Lomakin
Emmanuel Institute of Biochemical Physics, Russian Academy of Sciences, Moscow, Russia

Roman Makitra
Chemistry of Oxidizing Processes Division; Physical Chemistry of Combustible Minerals Department; Institute of Physical–Organic Chemistry & Coal Chemistry named after L. M. Lytvynenko, National Academy of Science of Ukraine 79053, Ukraine; Email: bazyljak.L.I@nas.gov.ua

Halyna Midyana
Chemistry of Oxidizing Processes Division; Physical Chemistry of Combustible Minerals Department, Institute of Physical–Organic Chemistry & Coal Chemistry named after L. M. Lytvynenko, National Academy of Science of Ukraine 79053, Ukraine, Email: bazyljak.L.I@nas.gov.ua

Vadim Z. Mingaleev
Institute of Organic Chemistry, Ufa Scientific Center of Russian Academy of Sciences, Ufa, Bashkortostan, 450054, Russia

S. M. Mousavi Motlagh
Department of Chemical Engineering, Imam Hossien University, Tehran, Iran

Olena Palchykova
Institute of Geology and Geochemistry of Combustible Minerals; National Academy of Science of Ukraine 79053, Ukraine

Anatolii A. Popov
Emmanuel Institute of Biochemical Physics, Russian Academy of Sciences, Moscow, Russia

S. Rafiei
University of Guilan, Rasht, Iran

Maziyar Sharifzadeh
Department of Chemical Engineering, Ayatollah Amoli Branch, Islamic Azad University, Amol, Iran

S. A. Vaziri
Department of Chemical Engineering, Imam Hossien University, Tehran, Iran

Genadiy E. Zaikov
N.M. Emanuel Institute of Biochemical Physics of Russian Academy of Sciences, Moscow 119991, Russia, Russian Federation

Vadim P. Zakharov
Bashkir State University, Ufa 450076, Bashkortostan, Russia, Email: zaharovvp@mail.ru

Elena M. Zakharova
Institute of Organic Chemistry, Ufa Scientific Center of Russian Academy of Sciences, Ufa, Bashkortostan, 450054, Russia

Azamat A. Zhansitov
Kabardino-Balkarian State University, Nalchik 360004, Russia, Russian Federation, Email: new_kompozit@mail.ru

LIST OF ABBREVIATIONS

ACs	active sites
CS	cuckoo search
DAC	dialdehyde cellulose
DAGA	diallylguanidine acetate
DAGTFA	diallylguanidine trifluoroacetate
DE	differential evolution
DLS	dynamic light scattering
DSS	dextran sulfate sodium salt
ESEM	environment scanning electron microscope
FPLC	fast protein liquid chromatography
MCC	microcrystalline cellulose
MWD	molecular weight distribution
NPBA	neutral polymeric bonding agent
OM	optical microscopy
PHB	poly(3-hydroxybutyrate)
PMMA	poly (methylmethacrylate)
PSO	particle swarm optimization
SA	sodium alginate
SALS	small angle light scattering
SC	sodium caseinate
SPEUs	segmented polyetherurethanes

LIST OF SYMBOLS

δ_1 and δ_2	Hildebrand's parameters
ρ_2	density of a polymer into solution
V_1	molar volume of the solvent
T_m	melting temperature
P_{max}	maximum pressure of each isotherm
S	number of point per isotherm per gas
C_{exp}	methane concentrations (experimental)
C_{cal}	methane concentrations (calculated)
U_e	electrophoretic mobility
ε	dielectric constant
η	viscosity
$z\rho$	zeta potential
k	Boltzmann's constant
T	temperature
n	dumbbells density
p	unit vector in nanoelement axis direction
ω_{ij}	rotation rate tensor
γ_{ij}	deformation tensor
D_r	rotary diffusivity
θ	shape factor
l	mean segment length
V	velocity
h	fabric thickness
m	equivalent mass
α	surface charge parameter

PREFACE

Polymers are substances that contain a large number of structural units joined by the same type of linkage. These substances often form into a chain-like structure. Starch, cellulose, and rubber all possess polymeric properties. Today, the polymer industry has grown to be larger than the aluminum, copper, and steel industries combined. Polymers already have a range of applications that far exceeds that of any other class of material available to man. Current applications extend from adhesives, coatings, foams, and packaging materials to textile and industrial fibers, elastomers, and structural plastics. Polymers are also used for most nanocomposites, electronic devices, biomedical devices, and optical devices, and are precursors for many newly developed high-tech ceramics.

This book presents leading-edge research in this rapidly changing and evolving field. Successful characterization of polymer systems is one of the most important objectives of today's experimental research of polymers. Considering the tremendous scientific, technological, and economic importance of polymeric materials, not only for today's applications but for the industry of the twenty-first century, it is impossible to overestimate the usefulness of experimental techniques in this field. Since the chemical, pharmaceutical, medical, and agricultural industries, as well as many others, depend on this progress to an enormous degree, it is critical to be as efficient, precise, and cost-effective in our empirical understanding of the performance of polymer systems as possible. This presupposes our proficiency with, and understanding of, the most widely used experimental methods and techniques. This book is designed to fulfill the requirements of scientists and engineers who wish to be able to carry out experimental research in polymers using modern methods.

Polymer nanocomposites are materials that possess unique properties. These properties are enhanced properties of the polymer matrix. Some of the improved properties are thermal stability, permeability to gases, flammability, mechanical strength and photodegradability. At complete dispersion of the new layers in the polymer matrix, these enhanced properties are obtained. The unique properties of the material makes it suitable in applications as, food and beverage packaging, automobile parts, furniture, carrier bags, electrical gadgets, and so on.

CHAPTER 1

ELECTROSPINNING PROCESS: A COMPREHENSIVE REVIEW AND UPDATE

S. RAFIEI

CONTENTS

1.1 INTRODUCTION

Understanding the nanoworld makes up one of the frontiers of modern science. One reason for this is that technology based on nanostructures promises to be hugely important economically [1–3]. Nanotechnology literally means any technology on a nanoscale that has applications in the real world. It includes the production and application of physical, chemical, and biological systems at scales ranging from individual atoms or molecules to submicron dimensions, as well as the integration of the resulting nanostructures into larger systems. Nanotechnology is likely to have a profound impact on our economy and society in the early twenty-first century, comparable with that of semiconductor technology, information technology, or cellular and molecular biology. Science and technology research in nanotechnology promises breakthroughs in areas such as materials and manufacturing [4], nanoelectronics [5], medicine and healthcare [6], energy [7], biotechnology [8], information technology [9], and national security [10]. It is widely felt that nanotechnology will be the next Industrial Revolution [9].

As far as "nanostructures" are concerned, one can view this as objects or structures whereby at least one of its dimensions is within nanoscale. A "nanoparticle" can be considered as a zero dimensional nanoelement, which is the simplest form of nanostructure. It follows that a "nanotube" or a "nanorod" is a one-dimensional nanoelement from which slightly more complex nanostructure can be constructed of Refs. [11–12].

Following this fact, a "nanoplatelet" or a "nanodisk" is a two-dimensional element which, along with its one-dimensional counterpart, is useful in the construction of nanodevices. The difference between a nanostructure and a nanodevice can be viewed upon as the analogy between a building and a machine (whether mechanical, electrical, or both) [1]. It is important to know that as far as nanoscale is concerned, these nanoelements should not be considered only as an element that forms a structure while they can be used as a significant part of a device. For example, the use of carbon nanotube as the tip of an atomic force microscope (AFM) would have it classified as a nanostructure. The same nanotube, however, can be used as a single-molecule circuit, or as part of a miniaturized electronic component, thereby appearing as a nanodevice. Hence, the function, along with the structure, is essential in classifiying which nanotechnology subarea it belongs to. This classification will be discussed in detail in further sections [11, 13].

As long as nanostructures clearly define the solids' overall dimensions, the same cannot be said so for nanomaterials. In some instances, a nanomaterial refers to a nanosized material; whereas in other instances, a nanomaterial is a bulk material with nanoscaled structures. Nanocrystals are other groups of nanostructured materials. It is understood that a crystal is highly structured and that the repetitive unit is indeed small enough. Hence, a nanocrystal refers to the size of the entire crystal itself being nanosized, but not of the repetitive unit [14].

Nanomagnetics are the other types of nanostructured materials that are known as highly miniaturized magnetic data storage materials with very high memory. This can be attained by taking advantage of the electron spin for memory storage; hence, the term "spin-electronics," which has since been more popularly and more conveniently known as "spintronics" [1, 9, 15]. In nanobioengineering, the novel properties of nanoscale are taken advantage of for bioengineering applications. The many naturally occurring nanofibrous and nanoporous structure in the human body further adds to the impetus for research and development in this subarea. Closely related to this is molecular functionalization whereby the surface of an object is modified by attaching certain molecules to enable desired functions to be carried out such as for sensing or filtering chemicals based on molecular affinity[16–17].

With the rapid growth of nanotechnology, nanomechanics are no longer the narrow field that it used to be[13]. This field can be broadly categorized into the molecular mechanics and the continuum mechanics approaches that view objects as consisting of discrete many-body system and continuous media, respectively. As long as the former inherently includes the size effect, it is a requirement for the latter to factor in the influence of increasing surface-to-volume ratio, molecular reorientation, and other novelties as the size shrinks. As with many other fields, nanotechnology includes nanoprocessing novel materials processing techniques by which nanoscale structures and devices are designed and constructed [18–19].

Depending on the final size and shape, a nanostructure or nanodevice can be created from the top-down or the bottom-up approach. The former refers to the act of removing or cutting down a bulk to the desired size; whereas, the latter takes on the philosophy of using the fundamental building blocks—such as atoms and molecules, to build up nanostructures in the same manner. It is obvious that the top-down and the bottom-up nanoprocessing methodologies are suitable for the larger and two smaller ends, respectively, in the spectrum of nanoscale construction. The effort of nanopatterning—or patterning at the nanoscale— would hence fall into nanoprocessing [1, 12, 18].

1.2   NANOSTRUCTURED MATERIALS

Strictly speaking, a nanostructure is any structure with one or more dimensions measuring in the nanometer (10^{-9}m) range. Various definitions refine this further, stating that a nanostructure should have a characteristic dimension lying between 1nm and 100 nm, putting nanostructures as intermediate in size between a molecule and a bacterium. Nanostructures are typically probed either optically (spectroscopy, photoluminescence ...), or in transport experiments. This field of investigation is often given the name mesoscopic transport, and the following considerations give an idea of the significance of this term[1–2, 12, 20–21].

What makes nanostructured materials very interesting and award them with their unique properties is that their size is smaller than critical lengths that characterize many physical phenomena. In general, physical properties of materials can be characterized by some critical length, a thermal diffusion length, or a scattering length, for example. The electrical conductivity of a metal is strongly determined by the distance that the electrons travel between collisions with the vibrating atoms or impurities of the solid. This distance is called the mean free path or the scattering length. If the sizes of the particles are less than these characteristic lengths, it is possible that new physics or chemistry may occur [1, 9, 17].

Several computational techniques have been used to simulate and model nanomaterials. Since the relaxation times can vary anywhere from picoseconds to hours, it becomes necessary to employ Langevin dynamics besides molecular dynamics in the calculations. Simulation of nanodevices through the optimization of various components and functions provides challenging and useful task[20, 22]. There are many examples where simulation and modeling have yielded impressive results, such as nanoscale lubrication [23]. Simulation of the molecular dynamics of DNA has been successful to some extent [24]. Quantum dots and nanotubes have been modeled satisfactorily [25–26]. First principles calculations of nanomaterials can be problematic if the clusters are too large to be treated by Hartree–Fock methods and too small for density functional theory [1]. In the next section various classifications of these kinds of materials are considered in detail.

1.2.1 NANOSTRUCTURED MATERIALS AND THEIR CLASSIFICATIONS

Nanostructure materials as a subject of nanotechnology are low-dimensional materials comprising building units of a submicron or nanoscale size at least in one direction and exhibiting size effects. The first classification idea of NSMs was given by Gleiter in 1995 [3]. A modified classification scheme for these materials, in which 0D, 1D, 2D, and 3D are included suggested in later researches [21] . These classifications are given below.

1.2.1.1 0D NANOPARTICLES

A major feature that distinguishes various types of nanostructures is their dimensionality. In the past 10 years, significant progress has been made in the field of 0D nanostructure materials. A rich variety of physical and chemical methods have been developed for fabricating these materials with well-controlled dimensions[3, 18]. Recently, 0D nanostructured materials such as uniform particles arrays (quantum dots), heterogeneous particles arrays, core-shell quantum dots, onions, hollow spheres, and nanolenses have been synthesized by several research groups[21]. They have been extensively studied in light-emitting diodes (LEDs), solar cells, single-electron transistors, and lasers.

1.2.1.2 1D NANOPARTICLES

In the past decade, 1D nanostructured materials have focused an increasing interest due to their importance in research and developments and have a wide range of potential applications [27]. It is generally accepted that these materials are ideal systems for exploring a large number of novel phenomena at the nanoscale and investigating the size and dimensionality dependence of functional properties. They are also expected to play a significant role as both interconnects and the key units in fabricating electronic, optoelectronic, and EEDs with nanoscale dimensions. The most important types of this group are nanowires, nanorods, nanotubes, nanobelts, nanoribbons, hierarchical nanostructures, and nanofibers [1, 18, 28].

1.2.1.3 2D NANOPARTICLES

2D nanostructures have two dimensions outside of the nanometric size range. In recent years, synthesis of 2D nanomaterial has become a focal area in materials research, owing to their many low-dimensional characteristics different from the bulk properties. Considerable research attention has been focused over the past few years on the development of them. Two-dimensional nanostructured materials with certain geometries exhibit unique shape-dependent characteristics and subsequent utilization as building blocks for the key components of nanodevices[21]. In addition, these materials are particularly interesting not only for basic understanding of the mechanism of nanostructure growth but also for investigation and developing novel applications in sensors, photocatalysts, nanocontainers, nanoreactors, and templates for 2D structures of other materials. Some of the 3D nanoparticles are junctions (continuous islands), branched structures, nanoprisms, nanoplates, nanosheets, nanowalls, and nanodisks [1].

1.2.1.4 3D NANOPARTICLES

Owing to the large specific surface area and other superior properties over their bulk counterparts arising from quantum size effect, they have attracted considerable research interest and many of them have been synthesized in the past 10 years [1, 12]. It is well known that the behaviors of NSMs strongly depend on the sizes, shapes, dimensionality and morphologies, which are thus the key factors to their ultimate performance and applications. Therefore, it is of great interest to synthesize 3D NSMs with a controlled structure and morphology. In addition, 3D nanostructures are an important material due to its wide range of applications in the area of catalysis, magnetic material and electrode material for batteries [2]. Moreover, the 3D NSMs have recently attracted intensive research interests because the nanostructures have higher surface area and supply enough absorption sites for all involved molecules in a small space [58]. On the contrary, such materials with porosity in three dimensions could lead to a better transport of the molecules. Nanoballs (dendritic structures), nanocoils, nanocones, nanopillers, and nanoflowers are in this group[1–2, 18, 29].

1.2.2 SYNTHESIS METHODS OF NANOMATERIALS

The synthesis of nanomaterials includes control of size, shape, and structure. Assembling the nanostructures into ordered arrays often becomes necessary for rendering them functional and operational. In the past decade, nanoparticles (powders) of ceramic materials have been produced in large scales by employing both physical and chemical methods. There has been considerable progress in the preparation of nanocrystals of metals, semiconductors, and magnetic materials by using colloid chemical methods [18, 30].

The construction of ordered arrays of nanostructures by using techniques of organic self-assembly provides alternative strategies for nanodevices. 2D and 3D arrays of nanocrystals of semiconductors, metals, and magnetic materials have been assembled by using suitable organic reagents [1, 31]. Strain directed assembly of nanoparticle arrays (e.g., of semiconductors) provides the means to introduce functionality into the substrate that is coupled to that on the surface[32].

Preparation of nanoparticles is an important branch of the materials science and engineering. The study of nanoparticles relates various scientific fields, for example, chemistry, physics, optics, electronics, magnetism, and mechanism of materials. Some nanoparticles have already reached practical stage. To meet the nanotechnology and nanomaterials development in the next century, it is necessary to review the preparation techniques of nanoparticles.

All particle synthesis techniques fall into one of the three categories: vapor-phase, solution precipitation, and solid-state processes. Although vapor-phase processes have been common during the early days of nanoparticles development, the last of the three processes mentioned above is the most widely used in the industry for production of micron-sized particles, predominantly due to cost considerations[18, 31].

Methods for preparation of nanoparticles can be divided into physical and chemical methods based on whether there exist chemical reactions [33]. On the contrary, in general, these methods can be classified into the gas-phase, liquid-phase, and solid-phase methods based on the state of the reaction system. The gas-phase method includes gas-phase evaporation method (resistance heating, high-frequency induction heating, plasma heating, electron beam heating, laser heating, electric heating evaporation method, vacuum deposition on the surface of flowing oil, and exploding wire method), chemical vapor reaction (heating heat pipe gas reaction, laser-induced chemical vapor reaction, plasma-enhanced chemical vapor reaction), chemical vapor condensation, and sputtering method. Liquid-phase method for synthesizing nanoparticles mainly includes precipitation, hydrolysis, spray, solvent thermal method (high temperature and high pressure), solvent evaporation pyrolysis, oxidation reduction (room pressure), emulsion, radiation chemical synthesis, and sol-gel processing. The solid-phase method includes thermal decomposition, solid-state reaction, spark discharge, stripping, and milling method [30, 33].

In other classification, there are two general approaches to the synthesis of nanomaterials and the fabrication of nanostructures, bottom-up and top-down approach. The first one includes the miniaturization of material components (up to atomic level) with further self-assembly process leading to the formation assembly of nanostructures. During self-assembly, the physical forces operating at nanoscale are used to combine basic units into larger stable structures. Typical examples are quantum dot formation during epitaxial growth and formation of nanoparticles from colloidal dispersion. The latter uses larger (macroscopic) initial structures, which can be externally controlled in the processing of nanostructures. Typical examples include etching through the mask, ball milling, and application of severe plastic deformation [3, 13]. Some of the most common methods are described in the sections that follow.

1.2.3 PLASMA-BASED METHODS

Metallic, semiconductive, and ceramic nanomaterials are widely synthesised by hot and cold plasma methods. A plasma is sometimes referred to as being "hot" if it is nearly fully ionized, or "cold" if only a small fraction, (e.g., 1%), of the gas molecules are ionized, but other definitions of the terms "hot plasma" and "cold plasma" are common. Even in cold plasma, the electron temperature is still typically several thousand degrees Celsius. In general, the related equipment consists of an arc-melting chamber and a collecting system. The thin films of alloys were prepared from highly pure metals by arc melting in an inert gas atmosphere. Each arc-melted ingot was flipped over and remelted three times. Then, the thin films of alloy were produced by arc melting a piece of bulk materials in a mixing gas atmosphere at a low pressure. Before the ultrafine particles were taken out from the arc-melting chamber, they were passivated with a mixture of inert gas and air to prevent the particles from burning up [34–35].

Cold plasma method is used for producing nanowires in large scale and bulk quantity. The general equipment of this method consists of a conventional horizontal quartz tube furnace and an inductively coupled coil driven by a 13.56 MHz radiofrequency (RF) power supply. This method often is called as an RF plasma method. During RF plasma method, the starting metal is contained in a pestle in an evacuated chamber. The metal is heated above its evaporation point using high-voltage RF coils wrapped around the evacuated system in the vicinity of the pestle. Helium gas is then allowed to enter the system, forming a high-temperature plasma in the region of the coils. The metal vapor nucleates on the He gas atoms and diffuses up to a colder collector rod where nanoparticles are formed. The particles are generally passivated by the introduction of some gas such as oxygen. In the case of aluminum nanoparticles, the oxygen forms a layer of aluminum oxide about the particle [1, 36].

1.2.4 CHEMICAL METHODS

Chemical methods have played a significant role in developing materials imparting technologically important properties through structuring the materials on the nanoscale. However, the primary advantage of chemical processing is its versatility in designing and synthesizing new materials that can be refined into the final end products. The secondary most advantage that the chemical processes offer over physical methods is a good chemical homogeneity, as a chemical method offers mixing at the molecular level. On the contrary, chemical methods frequently involve toxic reagents and solvents for the synthesis of nanostructured materials. Another disadvantage of the chemical methods is the unavoidable introduction of byproducts that require subsequent purification steps after the synthesis in other words, this process is time-consuming. Despite these facts, probably the most useful methods of synthesis in terms of their potential to be scaled up are chemical methods [33, 37]. There are a number of different chemical methods that can be used to make nanoparticles of metals, and we will give some examples. Several types of reducing agents can be used to produce nanoparticles such as $NaBEt_3H$, $LiBEt_3H$, and $NaBH_4$ where Et denotes the ethyl $(–C_2H_s)$ radical. For example, nanoparticles of molybdenum (Mo) can be reduced in toluene solution with $NaBEt_3H$ at room temperature, providing a high yield of Mo nanoparticles having dimensions of 1–5 nm [30].

1.2.4.1   THERMOLYSIS AND PYROLYSIS

Nanoparticles can be made by decomposing solids at high temperature having metal cations, and molecular anions or metal organic compounds. The process is called thermolysis. For example, small lithium particles can be made by decomposing lithium oxide, LiN_3. The material is placed in an evacuated quartz tube and heated to 400°C in the apparatus. At about 370°C, the LiN_3 decomposes, releasing N_2 gas, which is observed by an increase in the pressure on the vacuum gauge. In a few minutes, the pressure drops back to its original low value, indicating that all the N_2 has been removed. The remaining lithium atoms coalesce to form small colloidal metal particles. Particles less than 5 nm can be made by this method. Passivation can be achieved by introducing an appropriate gas [1].

Pyrolysis is commonly a solution process in which nanoparticles are directly deposited by spraying a solution on a heated substrate surface, where the constituent react to form a chemical compound. The chemical reactants are selected such that the products other than the desired compound are volatile at the temperature of deposition. This method represents a very simple and relatively cost-effective processing method (particularly, in regard to equipment costs) as compared to many other film deposition techniques [30].

The other pyrolysis-based method that can be applied in nanostructures production is a laser pyrolysis technique that requires the presence in the reaction medium of a molecule absorbing the CO_2 laser radiation [38–39]. In most cases, the atoms of a molecule are rapidly heated via vibrational excitation and are dissociated. But in some cases, a sensitizer gas such as SF_6 can be directly used. The heated gas molecules transfer their energy to the reaction medium by collisions leading to dissociation of the reactive medium without, in the ideal case, dissociation of this molecule. Rapid thermalization occurs after dissociation of the reactants due to transfer collision. Nucleation and growth of NSMs can take place in the as-formed supersaturated vapor. The nucleation and growth period is very short time (0.1–10 ms). Therefore, the growth is rapidly stopped as soon as the particles leave the reaction zone. The flame-excited luminescence is observed in the reaction region where the laser beam intersects the reactant gas stream. Since there is no interaction with any walls, the purity of the desired products is limited by the purity of the reactants. However, because of the very limited size of the reaction zone with a faster cooling rate, the powders obtained in this wellness reactor present a low degree of agglomeration. The particle size is small (~ 5–50 nm range) with a narrow size distribution. Moreover, the average size can be manipulated by optimizing the flow rate, and, therefore, the residence time in the reaction zone [39–40].

1.2.4.2   LASER-BASED METHODS

The most important laser-based techniques in the synthesis of nanoparticles are pulsed laser ablation. As a physical gas-phase method for preparing nanosized particles, pulsed laser ablation has become a popular method to prepare high-purity and ultrafine nanomaterials of any composition [41–42]. In this method, the material is evaporated using pulsed laser in a chamber filled with a known amount of a reagent gas and by controlling condensation of nanoparticles onto the support. It is possible to prepare nanoparticles of mixed molecular composition such as mixed oxides/nitrides and carbides/nitrides or mixtures of oxides of various metals by this method. This method is capable of a high rate of production of 2–3 g/min [40].

Laser chemical vapor deposition method is the next laser-based technique in which photoinduced processes are used to initiate the chemical reaction. During this method, three kinds of activation should be considered. First, if the thermalization of the laser energy is faster than the chemical reaction, pyrolytic, and/or photothermal activation is responsible for the activation. Second, if the first chemical reaction step is faster than the thermalization, photolytical (nonthermal) processes are responsible for the excitation energy. Third, combinations of the different types of activation are often encountered. During this technique, a high intensity laser beam is incident on a metal rod, causing evaporation of atoms from the surface of the metal. The atoms are then swept away by a burst of helium and passed through an orifice into a vacuum where the expansion of the gas causes cooling and formation of clusters of the metal atoms. These clusters are then ionized by UV radiation and passed into a mass spectrometer that measures their mass: charge ratio [1, 41–43].

Laser-produced nanoparticles have found many applications in medicine, biophotonics, in the development of sensors, new materials, and solar cells. Laser interactions provide a possibility of chemical clean synthesis, which is difficult to achieve under more conventional NP production conditions [42]. Moreover, a careful optimization of the experimental conditions can allow a control over size distributions of the produced nanoclusters. Therefore, many studies were focused on the investigation the laser nanofabrication. In particular, many experiments were performed to demonstrate nanoparticles formation in vacuum, in the presence of a gas or a liquid. Nevertheless, it is still difficult to control the properties of the produced particles. It is believed that numerical calculations can help explain experimental results and to better understand the mechanisms involved [43].

Despite rapid development in laser physics, one of the fundamental questions still concerns the definition of proper ablation mechanisms and the processes leading to the nanoparticles formation. Apparently, the progress in laser systems implies several important changes in these mechanisms, which depend on both laser parameters and material properties. Among the more studied ablation mechanisms there are thermal, photochemical and photomechanical ablation processes. Frequently, however, the mechanisms are mixed, so that the existing analytical equations are hardly applicable. Therefore, numerical simulation is needed to better understand and to optimize the ablation process [44].

Thus far, thermal models are commonly used to describe nanosecond (and longer) laser ablation. In these models, the laser-irradiated material experiences heating, melting, boiling, and evaporation. In this way, three numerical approaches were used [29, 45]:

Atomistic approach based on such methods as molecular dynamics (MD) and direct Monte Carlo (DSMC) simulation. Typical calculation results provide detailed information about atomic positions, velocities, kinetic, and potential energy.

Macroscopic approach based hydrodynamic models. These models allow the investigations of the role of the laser-induced pressure gradient, which is particularly important for ultra-short laser pulses. The models are based on a one fluid two-temperature approximation and a set of additional models (equation of state) that determines thermal properties of the target.

Multiscale approach based on the combination of two approaches cited above was developed by several groups and was shown to be particularly suitable for laser applications.

1.3 NANOFIBER TECHNOLOGY

Nanofiber consists of two terms "nano" and "fiber," as the latter term looks more familiar. Anatomists observed fibers as any of the filament constituting the extracellular matrix of connective tissue, or any elongated cells or thread-like structures, muscle fiber, or nerve fiber. According to textile industry, fiber is a natural or synthetic filament, such as cotton or nylon, capable of being spun into simply as materials made of such filaments. Physiologists and biochemists use the term "fiber" for indigestible plant matter consisting of polysaccharides such as cellulose, that when eaten stimulates intestinal peristalsis. Historically, the term "fiber" or "fibre" in British English comes from Latin "fibra." Fiber is a slender, elongated thread-like structure. Nano is originated from Greek word "nanos" or "nannos" refer to "little old man" or "dwarf." The prefixes "nannos" or "nano" as nannoplanktons or nanoplanktons used for very small planktons measuring 2–20 μm. In modern "nano" is used for describing various physical quantities within the scale of a billionth as nanometer (length), nanosecond (time), nanogram (weight), and nanofarad (charge) [1, 4, 9, 46]. As mentioned earlier, nanotechnology refers to the science and engineering concerning materials, structures, and devices, which has at least one dimension is 100nm or less. This term also refers for a fabrication technology, where molecules, specification, and individual atoms that have at least one dimension in nanometers or less is used to design or built objects. Nanofiber, as the name suggests, is the fiber having a diameter range in nanometer. Fibrous structure having at least 1D in nanometer or less is defined as nanofiber according to National Science Foundation (NSC). The term "nano" describes the diameter of the fibrous shape at anything below one micron or 1,000 nm [4, 18].

Nanofiber technology is a branch of nanotechnology whose primary objective is to create materials in the form of nanoscale fibers in order to achieve superior functions [1–2, 4]. The unique combination of high specific surface area, flexibility, and superior directional strength makes such fibers a preferred material form for many applications ranging from clothing to reinforcements for aerospace structures. Indeed, while the primary classification of nanofibers is that of nanostructure or nanomaterial, other aspects of nanofibers such as its characteristics, modeling, application, and processing would enable nanofibers to penetrate into many subfields of nanotechnology [4, 46–47].

It is obvious that nanofibers would geometrically fall into the category of 1D nanoscale elements that include nanotubes and nanorods. However, the flexible nature of nanofibers would align it along with other highly flexible nanoelements such as globular molecules (assumed as 0D soft matter), as well as solid and liquid films of nanothickness (2D). A nanofiber is a nanomaterial in view of its diameter, and can be considered a nanostructured material material if filled with nanoparticles to form composite nanofibers [1, 48].

The study of the nanofiber mechanical properties as a result of manufacturing techniques, constituent materials, processing parameters, and other factors would fall into the category of nanomechanics. Indeed, while the primary classification of nanofibers is that of nanostructure or nanomaterial, other aspects of nanofibers such as its characteristics, modeling, application, and processing would enable nanofibers to penetrate into many subfields of nanotechnology [1, 18].

Although the effect of fiber diameter on the performance and processibility of fibrous structures has long been recognized, the practical generation of fibers at the nanometer scale was not realized until the rediscovery and popularization of the electrospinning technology by Professor Darrell Reneker almost a decade ago [49–50]. The ability to create nanoscale fibers from a broad range of polymeric materials in a relatively simple manner using the electrospinning process, coupled with the rapid growth of nanotechnology in recent years have greatly accelerated the growth of nanofiber technology. Although there are several alternative methods available for generating fibers in a nanometer scale, none of the methods matches the popularity of the electrospinning technology due largely to the simplicity of the electrospinning process[18]. These methods will be discussed in the sections that follow.

1.3.1 *VARIOUS NANOFIBER PRODUCTION METHODS*

As was discussed in detail, nanofiber is defined as the fiber having at least 1D in nanometer range that can be used for a wide range of medical applications for drug delivery systems, scaffold formation, wound healing and widely used in tissue engineering, skeletal tissue, bone tissue, cartilage tissue, ligament tissue, blood vessel tissue, neural tissue, and so on. It is also used in dental and orthopedic implants [4, 51–52]. Nanofiber can be formed using different techniques including drawing, template synthesis, phases separation, self-assembly, and electrospinning.

1.3.1.1 *DRAWING*

In 1998, nanofibers were fabricated with citrate molecules through the process of drawing for the first time [53]. During drawing process, the fibers are fabricated by contacting a previously deposited polymer solution droplet with a sharp tip and drawing it as a liquid fiber that is then solidified by rapid evaporation of the solvent owing to the high surface area. The drawn fiber can be connected to another previously deposited polymer solution droplet, thereby forming a suspended fiber. Here, the predeposition of droplets significantly limits the ability to extend this technique, especially in 3D configurations and hard-to-access spatial geometries. Further, there is a specific time in which the fibers can be pulled. The viscosity of the droplet continuously increases with time due to solvent evaporation from the deposited droplet. The continual shrinkage in the volume of the polymer solution droplet affects the diameter of the fiber drawn and limits the continuous drawing of fibers [54].

To overcome the above-mentioned limitation, it is appropriate to use hollow glass micropipettes with a continuous polymer dosage. It provides greater flexibility in drawing continuous fibers in any configuration. Moreover, this method offers increased flexibility in the control of key parameters of drawing such as waiting time before drawing (because the required viscosity of the polymer edge drops), the drawing speed or viscosity, thus enabling repeatability and control on the dimensions of the fabricated fibers. Thus, drawing process requires a viscoelastic material that can undergo strong deformations while being cohesive enough to support the stresses developed during pulling [54–55].

1.3.1.2   TEMPLATE SYNTHESIS

Template synthesis implies the use of a template or mold to obtain a desired material or structure. Hence, the casting method and DNA replication can be considered as template-based synthesis. In the case of nanofiber creation by Feng et al. [56], the template refers to a metal oxide membrane with through-thickness pores of nanoscale diameter. Under the application of water pressure on the one side and restrain from the porous membrane causes extrusion of the polymer which, upon coming into contact with a solidifying solution, gives rise to nanofibers whose diameters are determined by the pores [1, 57].

This method is an effective route to synthesize nanofibrils and nanotubes of various polymers. The advantage of the template synthesis method is that the length and diameter of the polymer fibers and tubes can be controlled by the selected porous membrane, which results in more regular nanostructures. General feature of the conventional template method is that the membrane should be soluble so that it can be removed after synthesis so as to obtain single fibers or tubes. This restricts practical application of this method and gives rise to a need for other techniques [1, 56–57].

1.3.1.3   PHASE SEPARATION METHOD

This method consists of five basic steps: polymer dissolution, gelation, solvent extraction, freezing, and freeze-drying. In this process, it is observed that gelatin is the most difficult step to control the porous morphology of nanofiber. Duration of gelation varied with polymer concentration and gelation temperature. At low gelation temperature, nanoscale fiber network is formed; whereas, high gelation temperature led to the formation of platelet-like structure. Uniform nanofiber can be produced as the cooling rate is increased, polymer concentration affects the properties of nanofiber, as polymer concentration is increased porosity of fiber decreased and mechanical properties of fiber are increased [1, 58].

1.3.1.4   SELF-ASSEMBLY

Self-assembly refers to the build-up of nanoscale fibers using smaller molecules. In this technique, a small molecule is arranged in a concentric manner so that they can form bonds among the concentrically arranged small molecules that, upon extension in the plane-s normal, give the longitudinal axis of a nanofiber. The main mechanism for a generic self-assembly is the intramolecular forces that bring the smaller unit together. A hydrophobic core of alkyl residues and a hydrophilic exterior lined by peptide residues was found in obtained fiber. It is observed that the nanofibers produced with this technique have a diameter range of 5–8 mm approximately and are several microns in length [1, 59].

Although there are a number of techniques used for the synthesis of nanofiber, electrospinning represents an attractive technique to fabricate polymeric biomaterial into nanofibers. Electrospinning is one of the most commonly utilized methods for the production of nanofiber. It has a wide advantage over the previously available fiber formation techniques because here electrostatic force is used instead of conventionally used mechanical force for the formation of fibers. This method will be debated comprehensively in the sections that follow.

1.3.1.5   ELECTROSPINNING OF NANOFIBERS

Electrospinning is a straightforward and cost-effective method to produce novel fibers with diameters in the range of from less than 3 nm to over 1 mm, which overlaps contemporary textile fiber technology. During this process, an electrostatic force is applied to a polymeric solution to produce nanofiber [60–61] with diameter ranging from 50 to 1,000 nm or greater [49, 62–63]; Due to surface tension, the solution is held at the tip of syringe. Polymer solution is charged due to applied electric force. In the polymer solution, a force is induced due to mutual charge repulsion that is directly opposite to the surface tension of the polymer solution. Further increases in the electrical potential led to the elongation of the hemispherical surface of the solution at the tip of the syringe to form a conical shape known as "Taylor cone." [50, 64]. The electric potential is increased to overcome the surface tension forces to cause the formation of a jet, ejects from the tip of the Taylor cone. Due to elongation and solvent evaporation, charged jet instable and gradually thins in air primarily [62, 65–67]. The charged jet forms randomly oriented nanofibers that can be collected on a stationary or rotating grounded metallic collector [50]. Electrospinning provides a good method and a practical way of producing polymer fibers with diameters ranging from 40 to 2,000 nm [49–50].

1.3.1.5 1   THE HISTORY OF ELECTROSPINNING METHODOLOGY

William Gilbert discovered the first record of the electrostatic attraction of a liquid in 1,600 [68]. The first electrospinning patent was submitted by John Francis Cooley in 1900 [69]. After that in 1914, John Zeleny studied on the behavior of fluid droplets at the end of metal capillaries that caused the beginning of the mathematical model the behavior of fluids under electrostatic forces [65]. Between 1931 and 1944, Anton Formhals took out at least 22 patents on electrospinning [69]. In 1938, N.D. Rozenblum and I.V. Petryanov-Sokolov generated electrospun fibers, which they developed into filter materials [70]. Between 1964 and 1969, Sir Geoffrey Ingram Taylor produced the beginnings of a theoretical foundation of electrospinning by mathematically modeling the shape of the (Taylor) cone formed by the fluid droplet under the effect of an electric field [71–72]. In the early 1990s, several research groups (such as Reneker) demonstrated electrospun nanofibers. Since 1995, the number of publications about electrospinning has been increasing exponentially every year [69].

1.3.1.5 2   ELECTROSPINNING PROCESS

Electrospinning process can be explained in five significant steps including the folloiwng [48, 73–75]:

1. *Charging of the polymer fluid*: The syringe is filled with a polymer solution, the polymer solution is charged with a very high potential around 10–30 kV. The nature of the fluid and polarity of the applied potential free electrons, ions, or ion pairs are generated as the charge carriers form an electrical double layer. This charging induction is suitable for conducting fluid, but for nonconducting fluid charge directly injected into the fluid by the application of electrostatic field.
2. *Formation of the cone jet (Taylor cone)*: The polarity of the fluid depends on the voltage generator. The repulsion between the sim-

ilar charges at the free electrical double-layer works against the surface tension and fluid elasticity in the polymer solution to deform the droplet into a conical-shaped structure, that is known as a Taylor cone. Beyond a critical charge density Taylor cone becomes unstable and a jet of fluid is ejected from the tip of the cone.

3. *Thinning of the jet in the presence of an electric field:* The jet travels a path to the ground; this fluid jet forms a slender continuous liquid filament. The charged fluid is accelerated in the presence of an electrical field. This region of fluid is generally linear and thin.

4. *Instability of the jet:* Fluid elements accelerated under electric field and thus stretched and succumbed to one or more fluid instabilities that distort as they grow following many spiral and distort the path before collected on the collector electrode. This region of instability is also known as whipping region.

5. *Collection of the jet:* Charged electrospun fibers travel downfield until its impact with a lower potential collector plate. The orientation of the collector affects the alignment of the fibers. Different types of collector also affect the morphology and the properties of producing nanofiber. Different types of collectors are used—rotating drum collector, moving belt collector, rotating wheel with bevelled edge, multifilament thread, parallel bars, simple mesh collector, and so on.

1.3.1.5   3 ELECTROSPINNING SETUPS

Electrospinning is conducted at room temperature with atmospheric conditions. The typical setup of electrospinning apparatus is shown in Figure 1.1. Basically, an electrospinning system consists of three major components: a high-voltage power supply, a spinneret (such as a pipette tip), and a grounded collecting plate (usually a metal screen, plate, or rotating mandrel) and utilizes a high-voltage source to inject charge of a certain polarity into a polymer solution or melt, which is then accelerated toward a collector of opposite polarity [73, 76–77]. Most of the polymers are dissolved in some solvents before electrospinning, and when it completely dissolves, forms polymer solution. The polymer fluid is then introduced into the capillary tube for electrospinning. However, some polymers may emit unpleasant or even harmful smells; therefore, the processes should be conducted within chambers having a ventilation system. In the electrospinning process, a polymer solution held by its surface tension at the end of a capillary tube is subjected to an electric field and an electric charge is induced on the liquid surface due to this electric field. When the electric field applied reaches a critical value, the repulsive electrical forces overcome the surface tension forces. Eventually, a charged jet of the solution is ejected from the tip of the Taylor cone and an unstable and a rapid whipping of the jet occurs in the space between the capillary tip and collector, which leads to evaporation of the solvent, leaving a polymer behind. The jet is only stable at the tip of the spinneret and after that instability starts. Thus, the electrospinning process offers a simplified technique for fiber formation [50, 73, 78–79].

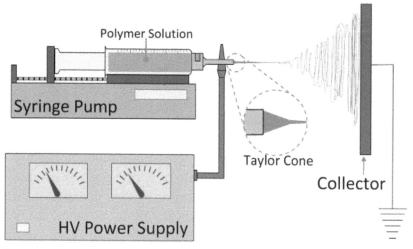

FIGURE 1.1   Scheme of a conventional electrospinning set-up.

1.3.1.5 4   THE EFFECTIVE PARAMETERS ON ELECTROSPINNING

The electrospinning process is generally governed by many parameters that can be classified broadly into solution parameters, process parameters, and ambient parameters. Each of these parameters significantly affects the fiber morphology obtained as a result of electrospinning; and by proper manipulation of these parameters, we can get nanofibers of desired morphology and diameters. These effective parameters are sorted as below [63, 67, 73, 76]: (a) Polymer solution parameters that includes molecular weight and solution viscosity, surface tension, solution conductivity, and dielectric effect of solvent and (b) processing parameters that include voltage, feed rate, temperature, effect of collector, and the diameter of the orifice of the needle.

(a) Polymer solution parameters

(1) Molecular weight and solution viscosity

The higher the molecular weight of the polymer increases molecular entanglement in the solution, the higher the increase in viscosity. The electrospun jet eject with high viscosity during it is stretched to a collector electrode leading to formation of continuous fiber with higher diameter, but very high viscosity makes difficult to pump the solution and also lead to the drying of the solution at the needle tip. As a very low viscosity lead in bead formation in the resultant electrospun fiber; therefore, the molecular weight and viscosity should be acceptable to form nanofiber [48, 80].

(2) Surface tension

Lower viscosity leads to decrease in surface tension resulting bead formation along the fiber length because the surface area is decreased, but at the higher viscosity effect of surface tension is nullified because of the uniform distribution of the polymer solution over the entangled polymer molecules. Therefore, lower surface tension is required to obtain smooth fiber and lower surface tension can be achieved by adding of surfactants in polymer solution [80–81].

(3) Solution conductivity

Higher conductivity of the solution followed a higher charge distribution on the electrospinning jet which leads to increase in stretching of the solution during fiber formation. Increased conductivity of the polymer solution lowers the critical voltage for the electrospinning. Increased charge leads to the higher bending instability leading to the higher deposition area of the fiber being formed, as a result jet path is increased and finer fiber is formed. Solution conductivity can be increased by the addition of salt or polyelectrolyte or increased by the addition of drugs and proteins that dissociate into ions when dissolved in the solvent formation of smaller diameter fiber [67, 80].

4) Dielectric effect of solvent

Higher the dielectric property of the solution lesser is the chance of bead formation and smaller is the diameter of electrospun fiber. As the dielectric property is increased, there is increase in the bending instability of the jet and the deposition area of the fiber is increased. As jet path length is increased fine fiber deposit on the collector [67, 80].

(b) Processing condition parameters

(1) Voltage

Taylor cone stability depends on the applied voltage; at higher voltage, greater amount of charge causes the jet to accelerate faster leading to smaller and unstable Taylor cone. Higher voltage leads to greater stretching of the solution due to fiber with small diameter formed. At lower voltage, the flight time of the fiber to a collector plate increases that led to the formation of fine fibers. There is greater tendency to bead formation at high voltage because of increased instability of the Taylor cone, and these beads join to form thick diameter fibers. It is observed that the better crystallinity in the fiber obtained at higher voltage. Instead of DC if AC voltage is provided for electrospinning, it forms thicker fibers [48, 80].

(2) Feed rate

As the feed rate is increased, there is an increase in the fiber diameter because greater volume of solution is drawn from the needle tip [80].

3) Temperature

At high temperature, the viscosity of the solution is decreased, and there is increase in higher evaporation rate that allows greater stretching of the solution and a uniform fiber is formed [82].

4) Effect of collector

In electrospinning, collector material should be conductive. The collector is grounded to create stable potential difference between needle and collector. A nonconducting material collector reduces the amount of fiber being deposited with lower packing density. But in case of conducting collector, there is accumulation of closely packed fibers with higher packing density. Porous collector yields fibers with lower packing density as compared with nonporous collector plate. In porous collector plate, the surface area is increased so residual solvent molecules gets evaporated fast as compared with nonporous. Rotating collector is useful in getting dry fibers as it provides more time to the solvents to evaporate. It also increases fibermorphology [83]. The specific hat target with proper parameters has a uniform surface electric field distribution, the target can collect the fiber mats of uniform thickness and thinner diameters with even-size distribution[80].

5) Diameter of pipette orifice

Orifice with small diameter reduces the clogging effect due to less exposure of the solution to the atmosphere and leads to the formation of fibers with smaller diameter. However, very small orifice has the disadvantage that it creates problems in extruding droplets of solution from the tip of the orifice [80].

1.4 DESIGN MULTIFUNCTIONAL PRODUCT BY NANOSTRUCTURES

The largest variety of efficient and elegant multifunctional materials is seen in natural biological systems, which occur sometimes in the simple geometrical forms in human-made materials. The multifunctionality of a material could be achieved by designing the material from the micro- to macroscales (bottom up design approach), mimicking the structural formations created by nature [84]. Biological materials present around us have a large number of ingenious solutions and serve as a source of inspiration. There are different ways of producing multifunctional materials that depend largely on whether these materials are structural composites, smart materials, or nanostructured materials. The nanostructure materials are most challenging and innovative processes, introducing, in the manufacturing, a new approaches such as self-assembly and self-replication. For biomaterials involved in surface-interface-related processes, common geometries involve capillaries, dendrites, hair, or fin-like attachments supported on larger substrates. It may be useful to incorporate similar hierarchical structures in the design and fabrication of multifunctional synthetic products that include surface sensitive functions such as sensing, reactivity, charge storage, transport property, or stress transfer. Significant effort is being directed to fabricate and understand materials that involve multiple-length scales and functionalities. Porous fibrous structures can behave like lightweight solids providing significantly higher surface area compared to compact ones. Depending on what is attached on their surfaces, or what matrix is infiltrated in them, these core structures can be envisioned in a wide variety of surface active components or net-shape composites. If nanoelements can be attached in the pores, the surface area within the given space can be increased by several orders of magnitude, thereby increasing the potency of any desired surface functionality. Recent developments in electrospinning have made these possible, thanks to a coelectrospinning polymer suspension [85]. This opens up the possibility of taking a functional material of any shape and size, and attaching nanoelements on them for added surface functionality. The fast-growing nanotechnology with modern computational/experimental methods gives the possibility to design multifunctional materials and products in human surroundings. Smart clothing, portable fuel cells, and medical devices are some of them. Research in nanotechnology began with applications outside of everyday life and is based on discoveries in physics and chemistry. The reason for that is need to understand the physical and chemical properties of molecules and nanostructures in order to control them. For example, nanoscale manipulation results in new functionalities for textile structures, including self-cleaning, sensing, actuating, and communicating. Development of precisely controlled or programmable medical nanomachines and nanorobots is great promise for nanomedicine. Once nanomachines are available, the ultimate dream of every medical man becomes reality. The miniaturization of instruments on micro- and nanodimensions promises to make our future lives safer with more humanity. A new approach in material synthesis is a computational-based material development. It is based on multiscale material and process modeling spanning, on a large spectrum of time as well as on length scales. Multiscale materials design means to design materials from a molecular scale up to a macroscale. The ability to manipulate at atomic and molecular level is also creating materials and structures that have unique functionalities and characteristics. Therefore it will be and revolutionizing next-generation technology ranging from structural materials to nanoelectro-mechanical systems (NEMs), for medicine and bioengineering applications. Recent research development in nanomaterials has been progressing at a tremendous speed for it can totally change the ways in which materials can be made with unusual properties. Such research includes the synthetic of nanomaterials, manufacturing processes, in terms of the controls of their nanostructural and geometrical properties, mouldability, and mixability with other matrix for nanocomposites. The cost of designing and producing a novel multifunctional material can be high and the risk of investment to be significant [12, 22].

Computational materials research that relies on multiscale modeling has the potential to significantly reduce development costs of new nanostructured materials for demanding applications by bringing physical and microstructural information into the realm of the design engineer. As there are various potential applications of nanotechnology in design multifunctional product, only some of the well-known properties come from by nanotreatment are critically highlighted [12, 22, 30]. This section reviews current research in nanotechnology application of the electrospinning nanofiber, from fibber production, and development to end uses as multifunctional nanostructure device and product. The electrospinning phenomena are described from experimental viewpoint to it simulation as multiscale problem.

1.4.1 THE MULTIFUNCTIONAL MATERIALS AND PRODUCTS

1.4.1.1 RESPONSIVE NANOPARTICLES

There are several directions in the research and development of the responsive nanoparticle (RNP) applications (Figure 1.2). Development of particles that respond by changing stability of colloidal dispersions is the first directions. Stimuli-responsive emulsions and foams could be very attractive for various technologies in coating industries, cosmetic, and personal care. The RNPs compete with surfactants; and hence, the costs for the particle production will play a key role. The main challenge is the development of robust and simple methods for the synthesis of RNPs from inexpensive colloidal particles and suspensions. That is indeed not a simple job since most of commercially available NPs are more expensive than surfactants. Another important application of RNPs for tunable colloidal stability of the particle suspensions is a very broad area of biosensors [86–87].

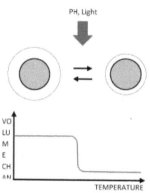

FIGURE 1.2   Stimuli-responsive nanoparticles.

The second direction is stimuli-responsive capsules that can release the cargo upon external stimuli. The capsules are interesting for biomedical applications (drugs delivery agents) and for composite materials (release of chemicals for self-healing). The most challenging task in many cases is to engineering systems capable to work with demanded stimuli. It is not a simple job for many biomedical applications where signalling biomolecules are present in very small concentrations and a range of changes of many properties is limited by physiological conditions. A well-known challenge is related to the acceptable size production of capsules. Many medical applications need capsules less than 50 nm in diameter. Fabrication of capsules with a narrow pore-size distribution and tunable sizes could dramatically improve the mass transport control [86, 88].

A hierarchically organized multicompartment RNPs are in the focus. These particles could respond to weak signals, multiple signals, and could demonstrate a multiple response. They can perform logical operations with multiple signals, store energy, absorb and consume chemicals, and synthesize and release chemicals. In other words, they could operate as an autonomous intelligent minidevice. The development of such RNPs can be considered as a part of biomimetics inspired by living cells or logic extension of the bottom up approach in nanotechnology. The development of the intelligent RNPs faces numerous challenges related to the coupling of many functional building blocks in a single hierarchically structured RNP. These particles could find applications for intelligent drug delivery, removal of toxic substances, diagnostics in medicine, intelligent catalysis, microreactors for chemical synthesis and biotechnology, new generation of smart products for personal use, and others [88–89].

1.4.1.2 NANOCOATINGS

In general, the coating's thickness is at least an order of magnitude lower than the size of the geometry to be coated. The coating's thickness less than 10 nm is called nanocoating. Nanocoatings are materials that are produced by shrinking the material at the molecular level to form a denser product. Nanostructure coatings have an excellent toughness, good corrosion resistance, wear, and adhesion properties. These coatings can be used to repair component parts instead of replacing them, resulting in significant reductions in maintenance costs. In addition, the nanostructure coatings will extend the service life of the component due to the improved properties over conventional coatings [90–91].

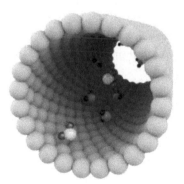

FIGURE 1.3 Nanocoatings.

1.4.1.3 FIBROUS NANOSTRUCTURE

The nanofibers are basic building block for plants and animals. From the structuralviewpoint, a uniaxial structure is able to transmit forces along its length and reducing required mass of materials. Nanofibers serve as another platform for multifunctional hierarchical example. The successful design concepts of nature, the nanofiber, become an attractive basic building component in the construction of hierarchically organized nanostructures. To follow nature's design, a process that is able to fabricate nanofiber from a variety of materials and mixtures becomes a prerequisite [92–93].

Control of the nanofibers arrangement is also necessary to optimize structural requirements. Finally, incorporation of other components into the nanofibers is required to form a complex, hierarchically organized composite. A nanofiber fabrication technique known as electrospinning process has the potential to play a vital role in the construction of a multilevels nanostructure [94]. In this chapter, we will introduce electrospinning as a potential technology for use as a platform for multifunctional, hierarchically organized nanostructures. Electrospinning is a method of producing superfine fibers with diameters ranging from 10 nm to 100 nm. Electrospinning occurs when the electrical forces at the surface of a polymer solution overcome the surface tension and cause an electrically charged jet of polymer solution to be ejected. A schematic drawing of the electrospinning process is shown in Figure 1.3. The electrically charged jet undergoes a series of electrically induced instabilities during its passage to the collection surface, which results in complicated stretching and looping of the jet [50, 60]. This stretching process is accompanied by the rapid evaporation of the solvent molecules, further reducing the jet diameter. Dry fibers are accumulated on the surface of the collector, resulting in a nonwoven mesh of nanofibers.

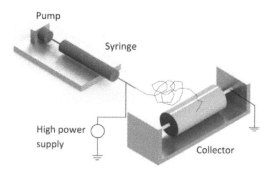

FIGURE 1.4   The electrospinning process.

Basically, an electrospinning system consists of three major components: a high-voltage power supply, an emitter (e.g., a syringe), and a grounded collecting plate (usually a metal screen, plate, or rotating mandrel). There are a wide range of polymers that used in electrospinning and are able to form fine nanofibers within the submicron range and used for varied applications. Electrospun nanofibers have been reported as being from various synthetic polymers, natural polymers or a blend of both including proteins, nucleic acids [74]. The electrospinning process is solely governed by many parameters, classified broadly into rheological, processing, and ambient parameters. Rheological parameters include viscosity, conductivity, molecular weight, and surface tension; whereas process parameters include applied electric field, tip to collector distance, and flow rate. Each of these parameters significantly affects the fibers morphology obtained as a result of electrospinning, and by proper manipulation of these parameters we can get nanofibers fabrics of desired structure and properties on multiple material scale (Figure 1.5).

Among these variables, ambient parameters encompass the humidity and temperature of the surroundings that play a significant role in determining the morphology and topology of electrospun fabrics. Nanofibrous assemblies such as nonwoven fibrous sheet, aligned fibrous fabric, continuous yarn, and 3D structure have been fabricated using electrospinning [51]. Physical characteristics of the electrospun nanofibers can also be manipulated by selecting the electrospinning conditions and solution. Structure organization on a few hierarchical levels (see Figure 1.4) has been developed using electrospinning. Such hierarchy and multifunctionality potential will be described in the sections that follow. Finally, we will describe how electrospun multifunctional, hierarchically organized nanostructure can be used in applications such as healthcare, defence and security, and environmental.

The slender-body approximation is widely used in electrospinning analysis of common fluids [51]. The presence of nanoelements (nanoparticles, carbon nanotube, clay) in suspension jet complicate replacement 3D axisymmetric with 1D equivalent jet problem under solid–fluid interaction force on nanolevel domain. The applied electric field induced dipole moment, while torque on the dipole rotate and align the nanoelement with electric field. The theories developed to describe the behavior of the suspension jet fall (Figure 1.6) into two levels: macroscopic and microscopic. The macroscopic-governing equations of the electrospinning are equation of continuity, conservation of the charge, balance of momentum, and electric field equation. Conservation of mass for the jet requires that [61, 95].

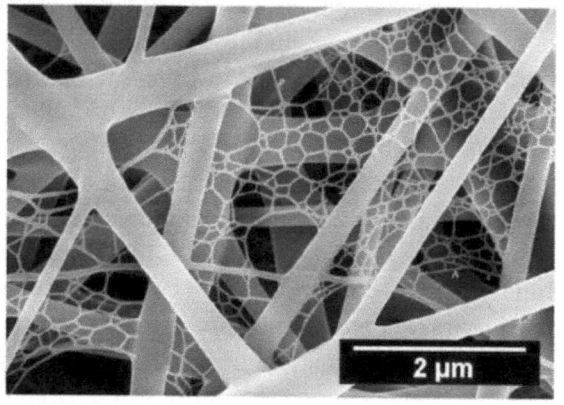

FIGURE 1.5   Multiscale electrospun fabric.

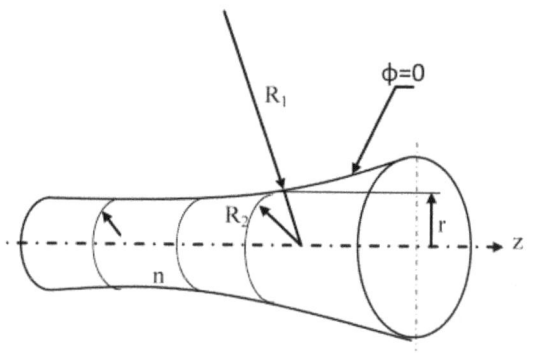

FIGURE 1.6   Geometry of the jet flow.

For polymer suspension stress tensor τ_{ij} come from polymeric $\hat{\tau}_{ij}$ and solvent contribution tensor via constitutive equation:

$$\tau_v = \hat{\tau}_v + n_s \cdot \dot{y}_v \qquad (1.4.1)$$

where η_s is solvent viscosity and γ_{ij} strain rate tensor. The polymer contribution tensor $\tau^\hat{}_{ij}$ depends on microscopic models of the suspension. Microscopic approach represents the microstructural features of material by means of a large number of micromechanical elements (beads, platelet, rods), obeying stochastic differential equations. The evolution equations of the microelements arise from a balance of momentum on the elementary level. For example, rheological behavior of the dilute suspension of the carbon nanotubes (CNTs) in polymer matrix can be described as FENE dumbbell model [96].

$$\lambda \langle Q.Q \rangle^{\nabla} = \delta_v - \frac{c \langle Q.Q \rangle}{1 - tr \langle Q.Q \rangle / b_{mAx}}$$
(1.4.2)>

where $\langle Q.Q \rangle$ is the suspension configuration tensor (see Figure 1.7), c is a spring constant, and max b is maximum CNT extensibility. Subscript ∇ represent the upper convected derivative, and λ denote a relaxation time. The polymeric stress can be obtained from the following relation:

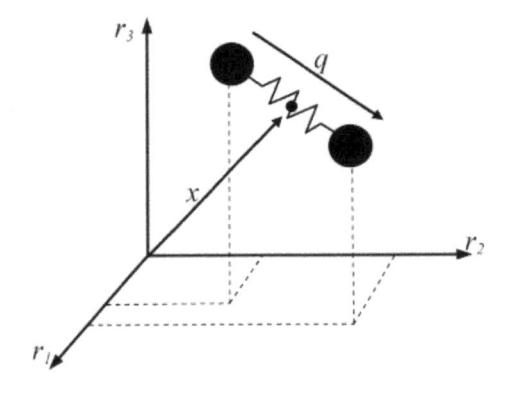

FIGURE 1.7   FENE dumbbell model.

$$\frac{\hat{\tau}_v}{nkT} = \delta_v - \frac{c \langle Q.Q \rangle}{1 - tr \langle Q.Q \rangle / b_{mAx}}$$
(1.4.3)

where k is Boltzmann's constant, T is temperature, and n is dumbbells density. Orientation probability distribution function ψ of the dumbbell vector Q can be described by the Fokker–Planck equation, neglecting rotary diffusivity.

$$\frac{\partial \psi}{\partial t} + \frac{\partial}{\partial Q} (\psi.Q) = 0$$
(1.4.4)

Solution equations (1.4.3) and (1.4.4), with supposition that flow in orifice is Hamel flow [97], give value orientation probability distribution function ψ along streamline of the jet. Rotation motion of a nanoelement (e.g., CNTs) in a Newtonian flow can be described as short fiber suspension model as another rheological model [8].

$$\frac{dp}{dt} = \frac{1}{2}\omega_v P_i + \frac{1}{2}\Theta\left[\frac{d\gamma_v}{dt}P_J - \frac{d\gamma_{kl}}{dt}P_k P_t P_l\right] - D_r\frac{1}{\psi}\frac{\partial\psi}{\partial t} \qquad (1.4.5)$$

where p is a unit vector in nanoelement axis direction, ω_{ij} is the rotation rate tensor, γ_{ij} is the deformation tensor, D_r is the rotary diffusivity, and θ is shape factor. Microscopic models for evolution of suspension microstructure can be coupled to macroscopic transport equations of mass and momentum to yield micro–macro multiscale flow models. The presence of the CNTs in the solution contributes to new form of instability with influences on the formation of the electrospun mat. The high strain rate on the nanoscale with complicated microstructure requires innovative research approach from the computational modeling viewpoint [98].

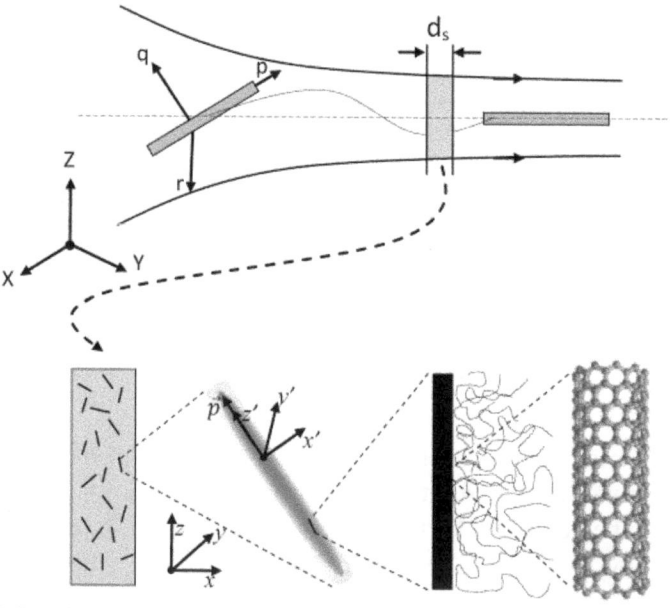

FIGURE 1.8   The CNTs alignment in jet flow.

By the illustrated multiscale treatment (Figure 1.8), the CNT suspension in the jet, one time as short flexible cylinder in solution (microscale), and second time as coarse grain system with polymer chain particles and CNT(nanoscale level).

1.5 MULTIFUNCTIONAL NANOFIBER-BASED STRUCTURE

The variety of materials and fibrous structures that can be electrospun allow for the incorporation and optimization of various functions to the nanofiber, either during spinning or through postspinning modifications. A schematic of the multilevel organization of an electrospun fiber based composite is shown in Figure 1.9.

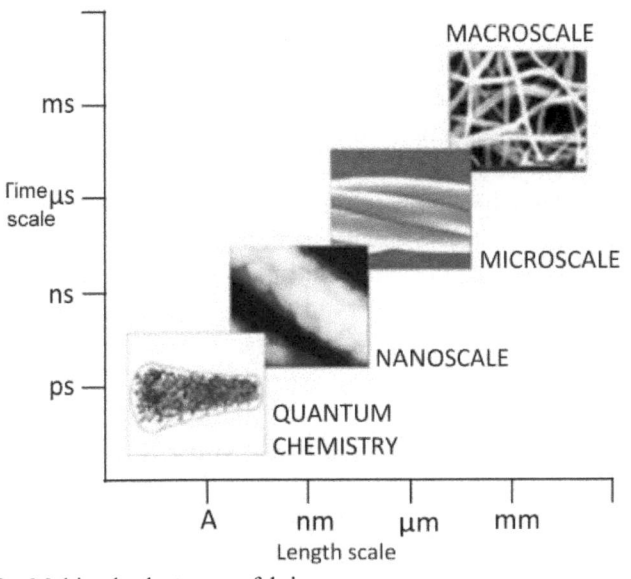

FIGURE 1.9 Multiscale electrospun fabrics.

Based on crrent technology, at least four different levels of organization can be put together to form a nanofiber-based hierarchically organized structure. At the first level, nanoparticles or a second polymer can be mixed into the primary polymer solution and electrospun to form composite nanofiber. Using a dual-orifice spinneret design, a second layer of material can be coated over an inner core material during electrospinning to give rise to the second-level organization. Two solution reservoirs, one leading to the inner orifice and the other to the outer orifice will extrude the solutions simultaneously. Otherwise, other conditions for electrospinning remain the same. Rapid evaporation of the solvents during the spinning process reduces mixing of the two solutions, thereby forming core-shell nanofiber. At the same level, various surface coating or functionalization techniques may be used to introduce additional property to the fabricated nanofiber surface. Chemical functionality is a vital component in advance multifunctional composite material to detect and respond to changes in its environment. Thus, various surface modifications techniques have been used to construct the preferred arrangement of chemically active molecules on the surface with the nanofiber as a supporting base. The third-level organization will see the fibers oriented and organized to optimize its performance. A multilayered nanofiber membrane or mixed materials nanofibers membrane can be fabricated *in situ* through selective spinning or using a multiple orifice spinneret design, respectively. Finally, the nanofibrous assembly may be embedded within a matrix to give the fourth-level organization. The resultant structure will have various properties and functionality due its hierarchical organization. Nanofiber structure at various levels have been constructed and tested for various applications and will be covered in the following sections. To follow surface functionality and modification, jet flow must be solved on multiple scale level. All above scale (nanoscale) can be solved by using particle method together with coarse-grain method on supramolecular level [50–51].

1.5.1 NANOFIBER EFFECTIVE PROPERTIES

The effective properties of the nanofiber can be determined by homogenization procedure using representative volume element (RVE). There is need for incorporating more physical information on microscale to precisely determine material behavior model. For electrospun suspension with nanoelements (CNTs), a concentric composite cylinder embedded with a caped carbon nanotube represents RVE as shown in Figure 1.10. A carbon nanotube with a length 2l, radii 2a is embedded at the center of matrix materials with a radii R and length 2L.

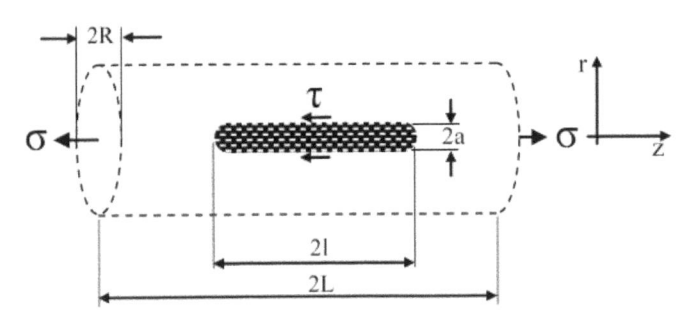

FIGURE 1.10   The nanofiber representative volume element.

The discrete atomic nanotube structure replaced the effective (solid) fiber having the same length and outer diameter as a discrete nanotube with effective Young's nanotube modulus determined from atomic structure. The stress and strain distribution in RVE was determined using modified shear-lag model [99]. For the known stress and strain distribution under RVE, elastic effective properties quantificators can be calculated. The effective axial module E_{33}, and the transverse module $E_{11} = E_{22}$, can be calculated as follow:

$$E_{33} = \frac{\langle \sigma_{zz} \rangle}{\langle \varepsilon_{zz} \rangle}$$
$$E_{11} = \frac{\langle \sigma_{xx} \rangle}{\langle \varepsilon_{xx} \rangle} \qquad (1.4.6)$$

where denotes a volume average under volume V as defined by

$$\langle \Xi \rangle = \frac{1}{V} \int_V \Xi\,(x, y, z).\, dV. \qquad (1.4.7)$$

The three-phase concentric cylindrical shell model has been proposed to predict effective modulus of nanotube reinforced nanofibers. The modulus of nanofiber depends strongly on the thickness of the interphase and CNTs diameter [12].

1.5.2   NETWORK MACROSCOPIC PROPERTIES

Macroscopic properties of the multifunctional structure determine final value of the any engineering product. The major objective in the determination of macroscopic properties is the link between atomic and continuum types of modeling and simulation approaches. The multiscale method such as quasicontinuum, bridge method, coarse-grain method, and dissipative particle dynamics are some popular methods of solution [98, 100]. The main advantage of the mesoscopic model is its higher computational efficiency than the molecular modeling without a loss of detailed properties at molecular level. Peridynamic modeling of fibrous network is another promising method, which allows damage, fracture, and long-range forces to be treated as natural components of the deformation of materials [101]. In the first stage, effective fiber properties are determined by homogenization procedure; whereas in the second stage, the point-bonded stochastic fibrous network at mesoscale is replaced by continuum plane stress model. Effective mechanical properties of nanofiber sheets at the macroscale level can be determined using the 2D Timoshenko beam network (figure 1.11). The critical parameters are the mean number of crossings per nanofiber, total nanofiber crossing in sheet and mean segment length [102]. Let us first consider a general planar fiber network characterized by fibre concentration n and fibre angular and length distribution ψ (ϕ, 1), where ϕ and l are fiber orientation angle and fiber length, respectively. The fibre radius r is considered uniform and the fibre concentration n is defined as the number of fiber per unite area.

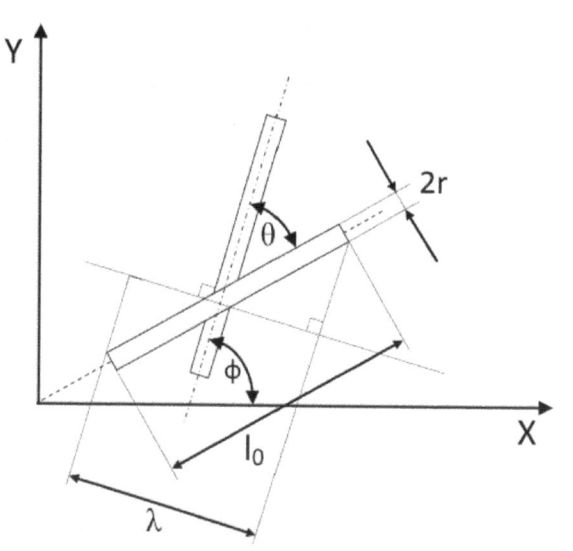

FIGURE 1.11   The fiber contact analysis.

The Poisson probability distribution can be applied to define the fiber segment length distribution for electrospun fabrics, a portion of the fiber between two neighboring contacts:

$$f(\ell) = \frac{1}{\ell}\exp\left(-\ell/\overline{\ell}\right) \tag{1.4.8}$$

where l is the mean segment length. The total number fiber segments $N^{\wedge}$ in the rectangular region b^* h:

$$\overline{N} = \{n.\ell_0(\langle\lambda\rangle + 2r) - 1\}.n.b.h \tag{1.4.9}$$

With

$$\langle\lambda\rangle = \int_0^\phi \int_0^\infty \psi(9.\ell).\lambda(9).d\ell.d9 \tag{1.4.10}$$

where the dangled segments at fiber ends have been excluded. The fiber network will be deformed in several ways. The strain energy in fiber segments comes from bending, stretching, and shearing modes of deformation can be calculated as follows (see Figure 1.12)

$$U = N.\ell_0.b.h\frac{1}{2}\iint \frac{E.A}{\ell}\varepsilon_{XX}^2.\psi(\varphi,\ell).\ell.d\ell.d\varphi$$

$$+n.\ell_0.\{(\langle\lambda\rangle + 2r) - 1\}.b.h.\frac{1}{2}\{\iint \frac{G.A}{\ell}..\gamma_{xy}^2.\psi(\varphi,\ell).\ell.d\ell.d\varphi \tag{1.4.11}$$

$$+ \iint \frac{3.E.I}{\ell^3}\gamma_{xy}^2.\psi(\varphi,\ell).\ell.d\ell.d\varphi \}$$

where A and I are beam cross-section area and moment of inertia, respectively. The first term on right-hand side is stretching mode, whereas the second and last terms are shear-bending modes, respectively.

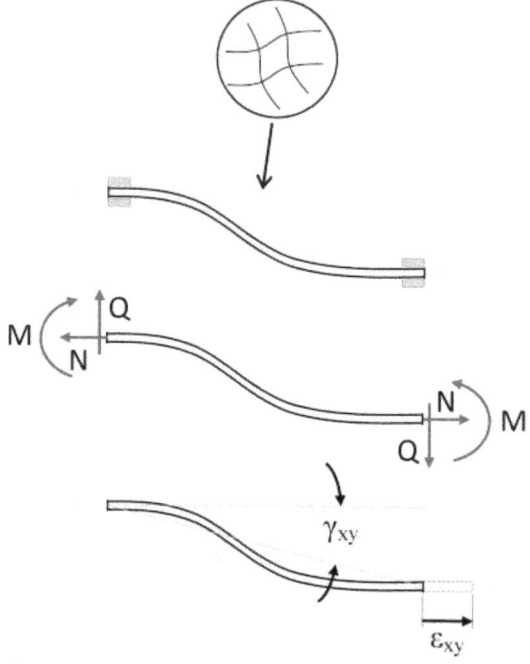

FIGURE 1.12   Fiber network 2D model.

The effective material constants for fiber network can be determined using homogenization procedure concept for fiber network. The strain energy fiber network for representative volume element is equal to strain energy continuum element with effective material constant. The strain energy of the representative volume element under plane stress conditions are as follows:

$$U = \frac{1}{2} \cdot \langle \varepsilon_v \rangle \cdot C_{vkl} \langle \varepsilon_{kl} \rangle \cdot V \qquad (1.4.12)$$

where is $V. b. h. 2. r$ representative volume element, C_{ijkl} are effective elasticity tensor. The square bracket $<>$ means macroscopic strain value. Microscopic deformation tensor was assume of a fiber segments ε_{ij} is compatible with effective macroscopic strain $\langle\varepsilon_{ij}\rangle$ of effective continuum (affine transformation). This is bridge relations between fiber segment microstrain ε_{ij} and macroscopic strain $<\varepsilon_{ij}>$ in the effective medium. Properties of this nanofibrous structure on the macro scale depend on the 3D joint morphology. The joints can be modeled as contact torsional elements with spring and dashpot [102]. The elastic energy of the whole random fiber network can be calculated numerically, from the local deformation state of the each segment by finite element method [103]. The elastic energy of the network is then the sum of the elastic energies of all segments. We consider here tensile stress, and the fibers are rigidly bonded to each other at every fiber–fiber crossing points. To mimic the microstructure of electrospun mats, we generated fibrous structures with fibers positioned in horizontal planes, and stacked the planes on top of one another to form a 2D or 3D structure. The representative volume element dimensions are considered to be an input parameter that can be used among other parameters to control the solid volume fraction of the structure, density number of fiber in the simulations. The number of intersections/unit area and mean lengths are obtained from image analysis of electrospun sheets. For the random point field, the stochastic fiber network was generated. Using polar coordinates and having the centerline equation of each fiber, the relevant parameters confined in the simulation box is obtained. The procedure is repeated until reaching the desired parameters is achieved [22, 95, 104]. The nonload-bearing fiber segments were removed and trimmed to keep dimensions $b*h$ of the representative window (see Figure 1.13).

A line representative network model is replaced by finite element beam mesh. The finite element analyses were performed in a network of 100 fibers, for some CNTs volume fractions values. Nanofibers were modeled as equivalent cylindrical beam as mentioned above. Effective mechanical properties of nanofiber sheets at the macroscale level can be determined using the 2D Timoshenko beam network.

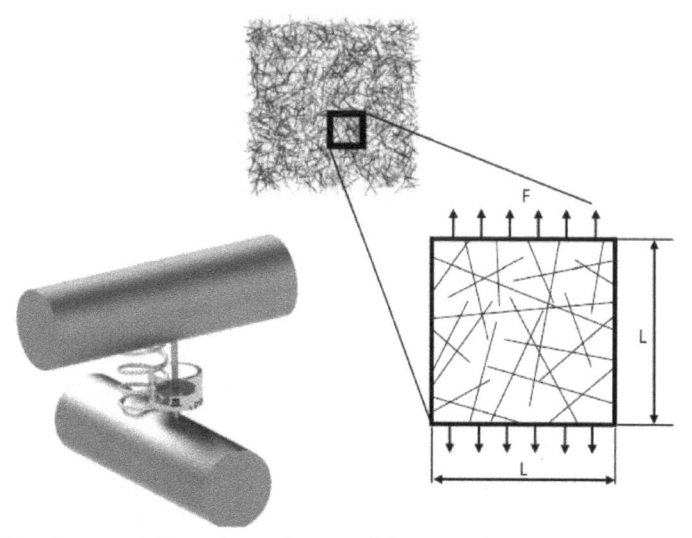

FIGURE 1.13   Repesentative volume element of the network.

For a displacement-based form of beam element, the principle of virtual work is assumed valid. For a beam system, a necessary and sufficient condition for equilibrium is that the virtual work done by sum of the external forces and internal forces vanish for any virtual displacement $\delta W = 0$. *The W* is the virtual work that the work is done by imaginary or virtual displacements.

$$\delta W = \oiiint_{v} v \delta \varepsilon_{v} + \oiiint_{v} F_{j} \delta u_{s} dV + \oiint_{A} T \delta u_{l} dA \qquad (1.4.13)$$

where ε is the strain, σ is the stress, F is the body force, δu is the virtual displacement, and T is the traction on surface A. The symbol δ is the variational operator designating the virtual quantity. Finite element interpolation for displacement field [15]:

$$\{u\} = [N]\{\hat{u}\} \qquad (1.4.14)\!\!>$$

where is $\{u\}$ u displacement vector of arbitrary point and $\{u\hat{}\}$ is nodal displacement point's vector. $[N]$ is shape function matrix. After FEM procedure, the problem is reduced to the solution of the linear system of equations

$$[K_e].\{u\} = \{f\} \tag{1.4.15}$$

where $\{u\}$ is global displacement vector, $\{f\}$ is the global nodal force vector, and $[K_e]$ is the global stiffness matrix. Finite element analyses were performed for computer generated network of 100 fibers. The comparison of calculated data with experimental data [99] for nanotube sheet shows some discrepancies (Figure 1.14). A rough morphological network model for the sheets can explain this on the one hand and simple joint morphology on the other hand [103].

$$\{u\} = [N]\{\hat{u}\}$$

FIGURE 1.14   The stress–strain curve.

1.5.3   FLOW IN FIBER NETWORK

Electrospun nanofiber materials are becoming an integral part of many recent applications and products. Such materials are currently being used in tissue engineering, air and liquid filtration, protective clothing's, drug delivery, and many others. Permeability of fibrous media is important in many applications, therefore during the past few decades, there have been many original studies dedicated to this subject. Depending on the fiber diameter and the air thermal conditions, there are four different regimes of flow around a fiber:

a)   Continuum regime ($K_N \leq 10^{-3}$),
b)   Slip-flow regime ($10^{-3} \leq K_N \leq 0.25$),
c)   Transient regime ($0.25 \leq K_N \leq 10$).
d)   Free molecule regime ($N_K \geq 10$),

Here, $K_N = 2\lambda /d$ is the fiber Knudson number, where $\lambda = RT//\sqrt{2}N^{\wedge}\pi$. d^2p is the mean free path of gas molecules, d is fiber diameter, and $N^{\wedge}$ is Avogadro number. Air flow around most electrospun nanofibers is typically in the slip or transition flow regimes. In the context of air filtration, the 2D analytical work of Kuwabara [105] has long been used for predicting the permeability across fibrous filters. The analytical expression has been modified by Brown [106] to develop an expression for predicting the permeability across filter media operating in the slip flow regime. The ratio of the slip to no-slip pressure drops obtained from the simplified 2D models may be used to modify the more realistic, and so more accurate, existing 3D permeability models in such a way that they could be used to predict the permeability of nanofiber structure. To test this supposition, for above developed 3D virtual nanofibrous structure, the Stokes flow equations solved numerically inside these virtual structures with an appropriate slip boundary condition that is developed for accounting the gas slip at fiber surface.

1.5.3.1   FLOW FIELD CALCULATION

A steady-state, laminar, incompressible model has been adopted for the flow regime inside our virtual media. Implemented in the Fluent code is used to solve continuity and conservation of linear momentum in the absence of inertial effects [107]:

$$\nabla . v = 0 \qquad (1.4.16)$$

$$\nabla P = \mu . \Delta^2 v \qquad (1.4.17)$$

The grid size required to mesh the gap between two fibers around their crossover point is often too small. The computational grid used for computational fluid dynamics (CFD) simulations needs to be fine enough to resolve the flow field in the narrow gaps, and at the same time coarse enough to cover the whole domain without requiring infinite computational power. Permeability of a fibrous material is often presented as a function of fiber radius, r, and solid volume fraction α, of the medium. Here, we use the continuum regime analytical expressions of Jackson and James [108], developed for 3D isotropic fibrous structures given as follows:

$$\frac{k}{r^2} = \frac{3r^2}{20a}[-\ell n(a) - 0.931] \qquad (1.4.18)$$

Brown [106] has proposed an expression for the pressure drop across a fibrous medium based on the 2D cell model of Kuwabara[105] with the slip boundary condition:

$$\Delta P_{SLIP} = \frac{4\mu a. hV. (1 + 1.996K_N}{r^2[\hat{K} + 1.996K_N(-0.5.\ell na - 0.25 + 0.25a^2]} \quad (1.4.19)$$

$$\hat{K} = -0.5.\ell na - 0.25 + 0.25a^2$$

where $K^\wedge = -0.5.\, ln\alpha - 0.75 + \alpha - 0.25\alpha^2$, Kuwabara hydrodynamic factor, h is fabric thickness, and V is velocity. As discussed in the some reference [101-108], permeability (or pressure drop) models obtained using ordered 2D fiber arrangements are known for underpredicting the permeability of a fibrous medium. To overcome this problem, if a correction factor can be derived based on the above 2D expression, and used with the realistic expressions developed for realistic 3D fibrous structures. From Eq. (1.4.19), we have for the case of no-slip boundary condition ($K_N = 0$):

$$\Delta P_{NONSLIP} = \frac{4\mu a.hV}{r^2\hat{K}} \quad (1.4.20)$$

The correction factor is defined as follows:

$$\Xi = \frac{\Delta P_{NONSLIP}}{\Delta P_{SLIP}} \quad (1.4.21)$$

to be used in modifying the original permeability expressions of Jackson and James [109], and/or any other expression based on the no-slip boundary condition, in order to incorporate the slip effect. For instance, the modified expression of Jackson and James can be presented as follows:

$$k_z = \frac{3r^2}{20\alpha}[-\ln(\alpha) - 0.931].\Xi \quad (1.4.22)$$

Operating pressure has no influence on the pressure drop in the continuum region, whereas pressure drop in the free molecular region is linearly proportional to the operating pressure. Although there are many equations available for predicting the permeability of fibrous materials made up of coarse fibers, there are no accurate "easy-to-use" permeability expressions that can be used for nanofiber media. On Figure 1.15 are drown corrected Jackson and James data (blue line). Points on figure are CFD numerical data.

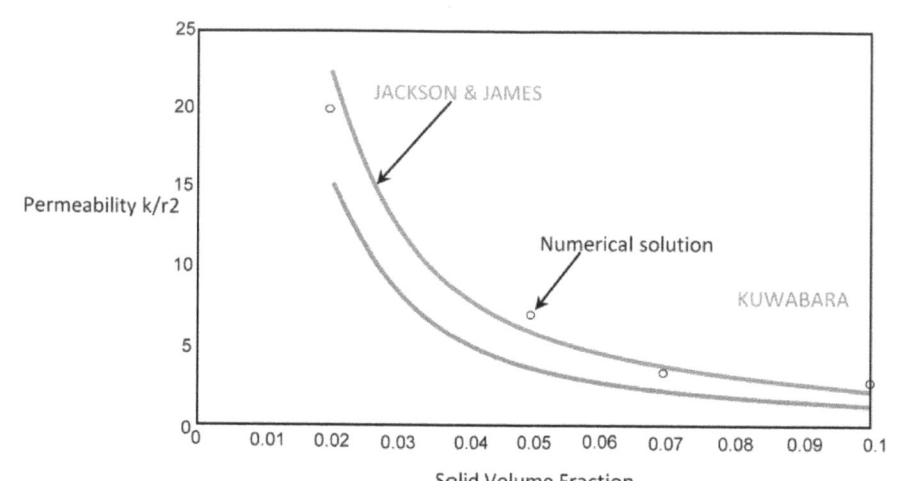

FIGURE 1.15　Permeability k/r^2 dependence on solid volume fraction.

1.5.4　SOME ILLUSTRATIVE EXAMPLES

1.5.4.1　FUEL CELL EXAMPLE

Fuel cells are electrochemical devices capable of converting hydrogen or hydrogen-rich fuels into electrical current by a metal catalyst. There are many kinds of fuel cells, such as proton exchange mat (PEM) fuel cells, direct methanol fuel cells, alkaline fuel cells, and solid oxide fuel cells [110]. PEM fuel cells are the most important ones among them because of high power density and low operating temperature. Pt nanoparticle catalyst is a main component in fuel cells. The price of Pt has driven up the cell cost and limited the commercialization. Electrospun materials have been prepared as alternative catalyst with high catalytic efficiency, good durability, and affordable cost. Binary PtRh and PtRu nanowires were synthesized by electrospinning, and they had better catalytic performance than commercial nanoparticle catalyst because of the one-dimensional features [111]. Pt nanowires also showed higher catalytic activities in a polymer electrolyte membrane fuel cell [112]. Instead of direct use as catalyst, catalyst supporting material is another important application area for electrospun nanofibers. Pt clusters were electrodeposited on a carbon nanofiber mat for methanol oxidation, and the catalytic peak current of the composite catalyst reached 420 mA/mg compared with 185 mA/mg of a commercial

Pt catalyst [113]. Pt nanoparticles were immobilized on polyimide-based nanofibers using a hydrolysis process and Pt nanoparticles were also loaded on the carbon nanotube containing polyamic acid nanofibers to achieve high catalytic current with long-term stability [114]. Proton exchange mat is the essential element of PEM fuel cells and normally made of a Nafion film for proton conduction. Because pure Nafion is not suitable for electrospinning due to its low viscosity in solution, it is normally mixed with other polymers to make blend nanofibers. Blend Nafion/PEO nanofibers were embedded in an inert polymer matrix to make a proton conducting mat [115], and a high proton conductivity of 0.06–0.08 S/cm at 15 °C in water and low water swelling of 12-23 wt% at 25 °C were achieved [116].

1.5.4.2 PROTECTIVE CLOTHING EXAMPLE

The development of smart nanotextiles has the potential to revolutionize the functionality of our clothing and the fabrics in our surroundings (Figure 1.16). This is made possible by such developments as new materials, fibbers, and finishing; inherently conducting polymers; carbon nanotubes; an antimicrobial nanocoatings. These additional functionalities have numerous applications in healthcare, sports, military applications, fashion, and so on. Smart textiles become a critical part of the emerging area of body sensor networks incorporating sensing, actuation, control and wireless data transmission [51–52, 117].

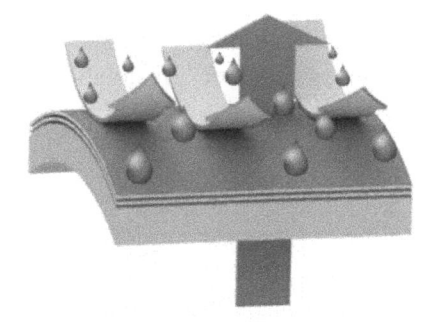

FIGURE 1.16 Ultrathin layer for selective transport.

1.5.4.3   MEDICAL DEVICE

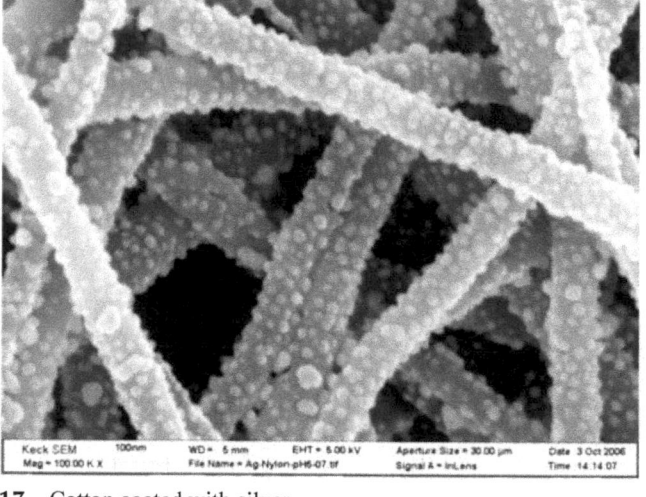

FIGURE 1.17   Cotton coated with silver.

Basic engineered nanomaterial and biotechnology products will enormously be useful in future medical applications. We know nanomedicine as the monitoring, repair, construction, and control of biological systems at the nanoscale level, using engineered nanodevices and nanostructures. The upper portion of the dress contains cotton coated with silver nanoparticles. Silver possesses (Figure 1.17) natural antibacterial qualities that are strengthened at the nanoscale, thus giving the ability to deactivate many harmful bacteria and viruses. The silver infusion also reduces the need to wash the garment, since it destroys bacteria, and the small size of the particles prevents soiling and stains [16, 118].

1.5.4.3.1   DRUG DELIVERY AND RELEASE CONTROL

Controlled release is an efficient process of delivering drugs in medical therapy. It can balance the delivery kinetics, immunize the toxicity and side effects, and improve patient convenience [119]. In a controlled-release system, the active substance is loaded into a carrier or device first, and then released at a predictable rate *in vivo* when administered by an injected or noninjected route. As a potential drug delivery carrier, electrospun nanofibers have exhibited many advantages. The drug loading is very easy to implement via electrospinning process, and the high applied voltage used in electrospinning process had little influence on the drug activity. The high specific surface area and short diffusion passage length give the nanofiber drug system higher overall release rate than the bulk material (e.g., film). The release profile can be finely controlled by modulation of nanofiber morphology, porosity, and composition.

Nanofibers for drug release systems mainly come from biodegradable polymers and hydrophilic polymers. Model drugs that have been studied include water soluble[120], poor water-soluble[121], and water insoluble drugs [122]. The release of macromolecules, such as DNA [123] and bioactive proteins from nanofibers was also investigated. In most cases, water-soluble drugs, including DNA and proteins, exhibited an early-stage burst [124]. For some applications, preventing postsurgery induced adhesion for instance, and such an early burst release will be an ideal profile because most infections occur within the first few hours after surgery. A recent study also found that when a poorly water soluble drug was loaded into PVP nanofibers [125], 84.9% of the drug can be released in the first 20 seconds when the drug-to-PVP ratio was kept as 1:4, which can be used for fast drug delivery systems. However, for a long-lasting release process, it would be essential to maintain the release at an even and stable pace, and any early burst release should be avoided. For a water insoluble drug, the drug release from hydrophobic nanofibers into buffer solution is difficult. However, when an enzyme capable of degrading nanofibers exists in the buffer solution, the drug can be released at a constant rate because of the degradation of nanofibers [122]. For example, when rifampin was encapsulated in PLA nanofibers, no drug release was detected from the nanofibers. However, when the buffer solution contained proteinase K, the drug release took place nearly in zero-order kinetics, and no early burst release happened. Similarly, initial burst release did not occur for poor water-soluble drugs, but the release from a nonbiodegradable nanofiber could follow different kinetics [126]. In another example, blending a hydrophilic but water-insoluble polymer (PEG-g-CHN) with PLGA could assist in the release of a poor water-soluble drug Iburprofen [127]. However, when a water soluble polymer was used, the poorly soluble drug was released accompanied with dissolving of the nanofibers, leading to a low burst release [128]. The early burst release can be reduced when the drug is encapsulated within the nanofiber matrix. When an amphiphilic block copolymer, PEG-b-PLA was added into Mefoxin/PLGA nanofibers, the cumulative amount of the released drug at earlier time points was reduced and the drug release was prolonged [129]. The reason for the reduced burst release was attributed to the encapsulation of some drug molecules within the hydrophilic block of the PEG-b-PLA. Amphiphilic block copolymer also assisted the dispersion and encapsulation of water-soluble drug into nanofibers when the polymer solution used an oleophilic solvent, such as chloroform, during electrospinning [130]. In this case, a water-in-oil emulation can be electrospun into uniform nanofibers, and drug molecules are trapped by hydrophilic chains. The swelling of the hydrophilic chains during releasing assists the diffusion of drug from nanofibers to the buffer. Coating nanofibers with a shell could be an effective way to control the release profile.

When a thin layer of hydrophobic polymer, such as poly (p-xylylene) (PPX), was coated on PVA nanofibers loaded with bovine serum albumin (BSA)/luciferase, the early burst release of the enzyme was prevented [131]. Fluorination treatment [132] on PVA nanofibers introduced functional C-F groups and made the fiber surface hydrophobic, which dramatically decreased the initial drug burst and prolonged the total release time. The polymer shell can also be directly applied via a coaxial coelectrospinning process, and the nanofibers produced are normally named "core-sheath" bicomponent nanofibers. In this case, even a pure drug can be entrapped into nanofiber as the core, and the release profile was less dependent on the solubility of drug released [133]. A research has compared the release behavior of two drug-loaded PLLA nanofibers prepared using blend and coaxial electrospinning [134]. It was found that the blend fibers still showed an early burst release, while the threads made of core-sheath fibers provided a stable release of growth factor and other therapeutic drugs. In addition, the early burst release can also be lowered via encapsulating drugs into nanomaterial, followed by incorporating the drug-loaded nanomaterials into nanofibers. For example, halloysite nanotubes loaded with tetracycline hydrochloride were incorporated into PLGA nanofibers and showed greatly reduced initial burst release [135].

1.6 CONCLUDING REMARKS OF MULTIFUNCTIONAL NANOSTRUCTURE DESIGN

Electrospinning is a simple, versatile, and cost-effective technology that generates nonwoven fibers with high surface area to volume ratio, porosity, and tunable porosity. Because of these properties, this process seems to be a promising candidate for various applications especially nanostructure applications. Electrospun fibers are increasingly being used in a variety of applications, such as tissue engineering scaffolds, wound healing, drug delivery, immobilization of enzymes, as membrane in biosensors, protective clothing, cosmetics, affinity membranes, filtration applications, and so on. In summary, mother nature has always used hierarchical structures such as capillaries and dendrites to increase multifunctional of living organs. Material scientists are at beginning to use this concept and create multiscale structures where nanotubes, nanofillers can be attached to larger surfaces and subsequently functionalized. In principle, many more applications can be envisioned and created. Despite several advantages and success of electrospinning there are some critical limitations in this process such as small pore size inside the fibers. Several attempts in these directions are being made to improve the design through multilayering, inclusion of nanoelements and blending with polymers with different degradation behavior. As new architectures develop, a new wave of surface-sensitive devices related to sensing, catalysis, photovoltaic, cell scaffolding, and gas storage applications is bound to follow.

1.7 INTRODUCTION TO THEORETICAL STUDY OF ELECTROSPINNING PROCESS

Electrospinning is a procedure in which an electrical charge to draw very fine (typically on the micro- or nanoscale) ibers from polymer solution or molten. Electrospinning shares characteristics of both and conventional solution dry spinning of fibers. The process does not require the use of coagulation chemistry or high temperatures to produce solid threads from solution. This makes the process more efficient to produce the fibers using large and complex molecules. Recently, various polymers have been successfully electrospun into ultrafine fibers mostly in solvent solution and some in melt form [79, 136]. Optimization of the alignment and morphology of the fibers is produced by fitting the composition of the solution and the configuration of the electrospinning apparatus such as voltage, flow rate, and and so on. As a result, the efficiency of this method can be improved [137]. Mathematical and theoretical modeling and simulating procedure will assist to offer an in-depth insight into the physical understanding of complex phenomena during electrospinningand might be very useful to manage contributing factors toward increasing production rate [75, 138].

Despite the simplicity of the electrospinning technology, industrial applications of it are still relatively rare, mainly due to the notable problems with very low fiber production rate and difficulties in controlling the process [67].

Modeling and simulation (M&S) give information about how something will act without actually testing it in real. The model is a representation of a real object or system of objects for purposes of visualizing its appearance or analyzing its behavior. Simulation is a transition from a mathematical or computational model for description of the system behavior based on sets of input parameters [104, 139]. Simulation is often the only means for accurately predicting the performance of the modeled system [140]. Using simulation is generally cheaper and safer than conducting experiments with a prototype of the final product. In addition, simulation can often be even more realistic than traditional experiments, as they allow the free configuration of environmental and operational parameters and can often be run faster than in real time. In a situation with different alternatives analysis, simulation can improve the efficiency, in particular when the necessary data to initialize can easily be obtained from operational data. Applying simulation adds decision support systems to the tool box of traditional decision support systems [141].

Simulation permits set up a coherent synthetic environment that allows for integration of systems in the early analysis phase for a virtual test environment in the final system. If managed correctly, the environment can be migrated from the development and test domain to the training and education domain in real systems under realistic constraints [142].

A collection of experimental data and their confrontation with simple physical models appears as an effective approach toward the development of practical tools for controlling and optimizing the electrospinning process. On the contrary, it is necessary to develop theoretical and numerical models of electrospinning because of demanding a different optimization procedure for each material[143]. Utilizing a model to express the effect of electrospinning parameters will assist researchers to make an easy and systematic way of presenting the influence of variables and by means of that, the process can be controlled. In addition, it causes to predict the results under a new combination of parameters. Therefore, without conducting any experiments, one can easily estimate features of the product under unknown conditions [95].

1.8 STUDY OF ELECTROSPINNING JET PATH

To yield individual fibers, most, if not all of the solvents must be evaporated by the time the electrospinning jet reaches the collection plate. As a result, volatile solvents are often used to dissolve the polymer. However, clogging of the polymer may occur when the solvent evaporates before the formation of the Taylor cone during the extrusion of the solution from several needles. To maintain a stable jet while still using a volatile solvent, an effective method is to use a gas jacket around the Taylor cone through two coaxial capillary tubes. The outer tube that surrounds the inner tube will provide a controlled flow of inert gas which is saturated with the solvent used to dissolve the polymer. The inner tube is then used to deliver the polymer solution. For 10 wt% poly (L-lactic acid) (PLLA) solution in dichloromethane, electrospinning was not possible due to clogging of the needle. However, when N2 gas was used to create a flowing gas jacket, a stable Taylor cone was formed and electrospinning was carried out smoothly.

1.8.1 THE THINNING JET (JET STABILITY)

The conical meniscus eventually gives rise to a slender jet that emerges from the apex of the meniscus and propagates downstream. Hohman et al. [60] first reported this approach for the relatively simple case of Newtonian fluids. This suggests that the shape of the thinning jet depends significantly on the evolution of the surface charge density and the local electric field. As the jet thins down and the charges relax to the surface of the jet, the charge density and local field quickly pass through a maximum, and the current due to advection of surface charge begins to dominate over that due to bulk conduction.

The crossover occurs on the length scale given by [6]:

$$L_N = \left(K^4 Q^7 \rho^3 (\ln X)^2 / 8\pi^2 E_\infty I^5 \varepsilon^{-2} \right)^{1/5}$$ (1.6.1)

This length scale defines the "nozzle regime" over which the transition from the meniscus to the steady jet occurs. Sufficiently far from the nozzle regime, the jet thins less rapidly and finally enters the asymptotic regime, where all forces except inertial and electrostatic forces cease to influence the jet. In this regime, the radius of the jet decreases as follows:

$$h = \left(\frac{Q^3 \rho}{2\pi^2 E_\infty I} \right)^{1/4} z^{-1/4}$$ (1.6.2)

Here, z is the distance along the centerline of the jet. Between the "nozzle regime" and the "asymptotic regime," the evolution of the diameter of the thinning jet can be affected by the viscous response of the fluid. Indeed by balancing the viscous and the electrostatic terms in the force balance equation, it can be shown that the diameter of the jet decreases:

$$h = \left(\frac{6\mu Q^2}{\pi E_\infty I} \right)^{1/2} z^{-1}$$ (1.6.3)

In fact, the straight jet section has been studied extensively to understand the influence of viscoelastic behavior on the axisymmetric instabilities [93] and crystallization [60] and has even been used to extract extensional viscosity of polymeric fluids at very high strain rates.

For highly strain-hardening fluids, Yu et al. [144] demonstrated that the diameter of the jet decreased with a power-law exponent of −1/2, rather than −1/4 or −1, as discussed earlier for Newtonian fluids. This −1/2 power-law scaling for jet thinning in viscoelastic fluids has been explained in terms of a balance between electromechanical stresses acting on the surface of the jet and the viscoelastic stress associated with extensional strain hardening of the fluid. In addition, theoretical studies of viscoelastic fluids predict a change in the shape of the jet due to non-Newtonian fluid behavior. Both Yu et al. [144] and Han et al. [145] have demonstrated that substantial elastic stresses can be accumulated in the fluid as a result of the highstrain rate in the transition from the meniscus into the jetting region.

This elastic stress stabilizes the jet against external perturbations. Further downstream the rate of stretching slows down, and the longitudinal stresses relax through viscoelastic processes. The relaxation of stresses following an extensional deformation, such as those encountered in electrospinning, has been studied in isolation for viscoelastic fluids [146]. Interestingly, Yu et al. [144] also observed that, elastic behavior notwithstanding, the straight jet transitions into the whipping region when the jet diameter becomes of the order of 10 mm.

1.8.2 THE WHIPPING JET (JET INSTABILITY)

While it is in principle possible to draw out the fibers of small diameter by electrospinning in the cone-jet mode alone, the jet does not typically solidify enough en route to the collector and succumbs to the effect of force imbalances that lead to one or more types of instability. These instabilities distort the jet as they grow. A family of these instabilities exists, and can be analyzed in the context of various symmetries (axisymmetric or nonaxisymmetric) of the growing perturbation to the jet.

Some of the lower modes of this instability observed in electrospinning have been discussed in a separate review [81]. The "whipping instability" occurs when the jet becomes convectively unstable and its centerline bends. In this region, small perturbations grow exponentially, and the jet is stretched out laterally. Shin et al. [62] and Fridrikh et al. [63] have demonstrated how the whipping instability can be largely responsible for the formation of solid fiber in electrospinning. This is significant, since as recently as the late 1990s, the bifurcation of the jet into two more or less equal parts (also called "splitting" or "splaying") were thought to be the mechanism through which the diameter of the jet is reduced, leading to the fine fibers formed in electrospinning. In contrast to "splitting" or "splaying," the appearance of secondary, smaller jets from the primary jet have been observed more frequently and in situ [64, 147]. These secondary jets occur when the conditions on the surface of the jet are such that perturbations in the local field, for example, due to the onset of the slight bending of the jet, is enough to overcome the surface tension forces and invoke a local jetting phenomenon.

The conditions necessary for the transition of the straight jet to the whipping jet has been discussed in the works of Ganan-Calvo [148], Yarin et al. [64], Reneker et al. [66], and Hohman et al. [60].

During this whipping instability, the surface charge repulsion, surface tension, and inertia were considered to have more influence on the jet path than Maxwell's stress, which arises due to the electric field and finite conductivity of the fluid. Using the equations reported by Hohman et al. [60] and Fridrikh et al. [63] obtained an equation for the lateral growth of the jet excursions arising from the whipping instability far from the onset and deep into the nonlinear regime. These developments have been summarized in the review article by Rutledge and Fridrikh.

The whipping instability is postulated to impose the stretch necessary to draw out the jet into fine fibers. As discussed earlier, the stretch imposed can make an elastic response in the fluid, especially if the fluid is polymeric in nature. An empirical rheological model was used to explore the consequences of nonlinear behavior of the fluid on the growth of the amplitude of the whipping instability in numerical calculations [63, 79]. There, it was observed that the elasticity of the fluid significantly reduces the amplitude of oscillation of the whipping jet. The elastic response also stabilizes the jet against the effect of surface tension. In the absence of any elasticity, the jet eventually breaks up and forms an aerosol. However, the presence of a polymer in the fluid can stop this breakup if

$$\tau / \left(\frac{\rho h^3}{\gamma} \right)^{1/2} \geq 1 \qquad (1.6.4)$$

where τ is the relaxation time of the polymer, ρ is the density of the fluid, h is a characteristic radius, and γ is the surface tension of the fluid.

1.9   ELECTROSPINNING DRAWBACKS

Electrospinning has attracted much attention both to academic research and industry application because electrospinning (1) can fabricate continuous fibers with diameters down to a few nanometers, (2) is applicable to a wide range of materials such as synthetic and natural polymers, metals as well as ceramics and composite systems, (3) and can prepare nanofibers with low cost and high yielding [47].

Despite the simplicity of the electrospinning technology, industrial applications of it are still relatively rare, mainly due to the notable problems of very low fiber production rate and difficulties in controlling the process [50, 67]. The usual feedrate for electrospinning is about 1.5ml/hr. Given a solution concentration of 0.2g/ml, the mass of nanofiber collected from a single needle after an hour is only 0.3g. In order for electrospinning to be commercially viable, it is necessary to increase the production rate of the nanofibers. To do so, multiple-spinning setup is necessary to increase the yield while at the same time maintaining the uniformity of the nanofiber mesh [48].

Optimization of the alignment and morphology of the fibers which is produced by fitting the composition of the solution and the configuration of the electrospinning apparatus such as voltage, flow rate, and so on, can be useful to improve the efficiency of this method [137]. Mathematical and theoretical modeling and simulating procedure will assist to offer an in-depth insight into the physical understanding of complex phenomena uring electrospinningand might be very useful to manage contributing factors toward increasing production rate [75, 138].

At present, nanofibers have attracted the attention of researchers due to their remarkable micro and nanostructural characteristics, high surface area, small pore size, and the possibility of their producing 3D structure that enables the development of advanced materials with sophisticated applications [73].

Controlling the property, geometry, and mass production of the nanofibers is essential to comprehend quantitatively how the electrospinning process transforms the fluid solution through a millimeter diameter capillary tube into solid fibers that are four to five orders smaller in diameter [74].

As mentioned earlier, the electrospinning gives us the impression of being a very simple and easily controlled technique for the production of nanofibers. But, in reality, the process is very intricate. Thus, electrospinning is usually described as the interaction of several physical instability processes. The bending and stretching of the jet are mainly caused by the rapidly whipping which is an essential element of the process induced by these instabilities. Until now, little is known about the detailed mechanisms of the instabilities and the splaying process of the primary jet into multiple filaments. It is thought to be responsible that the electrostatic forces overcome surface tensions of the droplet during undergoing the electrostatic field and the deposition of jets formed nanofibers [47].

Although electrospinning has been become an indispensable technique for generating functional nanostructures, many technical issues still need to be resolved. For example, it is not easy to prepare nanofibers with a same scale in diameters by electrospinning; it is still necessary to investigate systematically the correlation between the secondary structure of nanofiber and the processing parameters; the mechanical properties, photoelectric properties, and other property of single fiber should be systematically studied and optimized; the production of nanofiber is still in laboratory level, and it is extremely important to make efforts for scaled-up commercialization; nanofiber from electrospinning has a the low production rate and low mechanical strength which hindered its commercialization; in addition, another more important issue should be resolved is how to treat the solvents volatilized in the process.

Until now, lots of efforts have been put on the improvement of electrospinning installation, such as the shape of collectors, modified spinnerets, and so on. The application of multijets electrospinning provides a possibility to produce nanofibers in industrial scale. The development of equipments which can collect the poisonous solvents and the application of melt electrospinning, which would largely reduce the environment problem, create a possibility of the industrialization of electrospinning. The application of water as the solvent for elelctrospinning provide another approach to reduce environmental pollution, which is the main fact hindered the commercialization of electrospinning. In summary, electrospinning is an attractive and promising approach for the preparation of functional nanofibers due to its wide applicability to materials, low cost and high production rate [47].

1.10 MODELING THE ELECTROSPINNING PROCESS

The electrospinning process is a fluid dynamics-related problem. Controlling the property, geometry, and mass production of the nanofibers is essential to comprehend quantitatively how the electrospinning process transforms the fluid solution through a millimeter diameter capillary tube into solid fibers that are four to five orders smaller in diameter [74]. Although information on the effect of various processing parameters and constituent material properties can be obtained experimentally, theoretical models offer in-depth scientific understanding which can be useful to clarify the affecting factors that cannot be exactly measured experimentally. Results from modeling also explained how processing parameters and fluid behavior lead to the nanofiber of appropriate properties. The term "properties" refers to basic properties (i.e., fiber diameter, surface roughness, and fiber connectivity), physical properties (i.e, stiffness, toughness, thermal conductivity, electrical resistivity, thermal expansion coefficient, and density) and specialized properties (i.e., biocompatibility, degradation curve, and for biomedical applications) [48, 73].

For example, the developed models can be used for the analysis of mechanisms of jet deposition and alignment on various collecting devices in arbitrary electric fields [149].

Various methods formulated by researchers are prompted by several applications of nanofibers. It would be sufficient to briefly describe some of these methods to observed similarities and disadvantages of these approaches. An abbreviated literature review of these models will be discussed in the sections that follow.

1.10.1 MODELING ASSUMPTIONS

Just as in any other process modeling, a set of assumptions are required for the following reasons:

a. To furnish industry-based applications whereby speed of calculation, but not accuracy, is critical
b. To simplify, hence enabling checkpoints to be made before more detailed models can proceed
c. For enabling the formulations to be practically traceable

The first assumption to be considered as far as electrospinning is concerned is conceptualizing the jet itself. Even though the most appropriate view of a jet flow is that of a liquid continuum, the use of nodes connected in series by certain elements that constitute rheological properties has proven successful [64, 66]. The second assumption is the fluid constitutive properties. In the discrete node model [66], the nodes are connected in series by a Maxwell unit, that is, a spring and dashpot in series, for quantifying the viscoelastic properties.

In analyzing viscoelastic models, we apply two types of elements: the dashpot element that describes the force as being in proportion to the velocity (recall friction), and the spring element which describes the force as being in proportion to elongation. One can then develop viscoelastic models using combinations of these elements. Among all possible viscoelastic models, the Maxwell model was selected by Reneker et al. [66] due to its suitability for liquid jet as well as its simplicity. Other models are either unsuitable for liquid jet or too detailed.

In the continuum models a power law can be used for describing the liquid behavior under shear flow for describing the jet flow [150]. At this juncture, it can be noted that the power law is characterized from a shear flow, whereas the jet flow in electrospinning undergoes elongational flow. This assumption will be discussed in detail in the subsequent sections.

The other assumption that should be applied in electrospinning modeling is about the coordinate system. The method for coordinate system selection in electrospinning process is similar to other process modeling, the system that best bring out the results by (i) allowing the computation to be performed in the most convenient manner and, more importantly, (ii) enabling the computation results to be accurate. In view of the linear jet portion during the initial first stage of the jet, the spherical coordinate system is eliminated. Assuming the second stage of the jet to be perfectly spiraling, due to bending of the jet, the cylindrical coordinate system would be ideal. However, experimental results have shown that the bending instability portion of the jet is not perfectly expanding spiral. Hence, the Cartesian coordinate system, which is the most common of all coordinate system, is adopted.

Depending on the processing parameters (i.e., applied voltage and volume flow rate) and the fluid properties (i.e., surface tension and viscosity) as many as 10 modes of electrohydrodynamically driven liquid jet have been identified [151]. The scope of jet modes is highly abbreviated in this chapter because most electrospinning processes that lead to nanofibers consist of only two modes, the straight jet portion, and the spiraling (or whipping) jet portion. Insofar as electrospinning process modeling is involved, the following classification indicates the considered modes or portion of the electrospinning jet.

   1.   Modeling the criteria for jet initiation from the droplet [Senador et al. [152]; Yarin et al.[64]

2. Modeling the straight jet portion [153–154] Spivak et al. [150, 155]

3. Modeling the entire jet [Reneker et al. [66]; Yarin et al.[156]; Hohman et al. [60–61]

1.10.2 CONSERVATION RELATIONS

Balance of the producing accumulation is, particularly, a basic source of quantitative models of phenomena or processes. Differential balance equations are formulated for momentum, mass, and energy through the contribution of local rates of transport expressed by the principle of Newton's, Fick's, and Fourier laws. For a description of more complex systems such as electrospinning that involved strong turbulence of the fluid flow, characterization of the product property is necessary and various balances are required [157].

The basic principle used in modeling of chemical engineering process is a concept of balance of momentum, mass and energy, which can be expressed in a general form as follows:

$$A = I + G - O - C \qquad (1.8.1)$$

where
   A is the accumulation built up within the system.
   I is the input entering through the system surface.
   G is the generation produced in system volume.
   O is the output leaving through system boundary.
   C is consumption used in system volume.

The form of expression depends on the level of the process phenomenon description. [157–158]

According to the electrospinning models, the jet dynamics are governed by a set of three equations representing mass, energy, and momentum conservation for the electrically charge jet [159].

1.1.1. IN ELECTROSPINNING MODELING FOR SIMPLIFICATION OF DESCRIBING THE PROCESS, RESEARCHERS CONSIDER AN ELEMENT OF THE JET AND THE JET VARIATION VERSUS TIME IS NEGLECTED.

1.1.2. MASS CONSERVATION

The concept of mass conservation is widely used in many fields such as chemistry, mechanics, and fluid dynamics. Historically, mass conservation was discovered in chemical reactions by Antoine Lavoisier in the late eighteenth century, and was of decisive importance in the progress from alchemy to the modern natural science of chemistry. The concept of matter conservation is useful and sufficiently accurate for most chemical calculations, even in modern practice [160].

The equations for the jet follow from Newton's Law and the conservation laws obey, namely, conservation of mass and conservation of charge [60].

According to the conservation of mass equation

$$\pi R^2 v = Q \qquad (1.8.2)$$

$$\frac{\partial}{\partial t}\left(\pi R^2\right) + \frac{\partial}{\partial z}\left(\pi R^2 v\right) = 0 \qquad (1.8.3)$$

For incompressible jets, by increasing the velocity the radius of the jet decreases. At the maximum level of the velocity, the radius of the jet reduces. The macromolecules of the polymers are compacted together closer while the jet becomes thinner as shown in Figure 1.18. When the radius of the jet reaches the minimum value and its speed becomes maximum to keep the conservation of mass equation, the jet dilates by decreasing its density which is called electrospinning dilation [161–162].

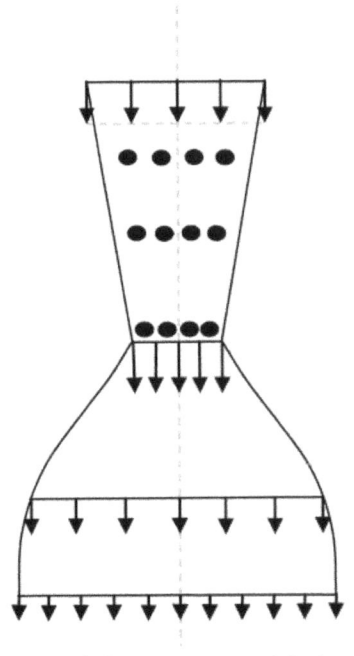

FIGURE 1.18   Macromolecular chains are compacted during the electrospinning.

1.10.3   ELECTRIC CHARGE CONSERVATION

An electric current is a flow of electric charge. Electric charge flows when there is voltage present across a conductor. In physics, charge conservation is the principle that electric charge can neither be created nor destroyed. The net quantity of electric charge, the amount of positive charge minus the amount of negative charge in the universe, is always conserved. The first written statement of the principle was by American scientist and statesman Benjamin Franklin in 1747 [163]. Charge conservation is a physical law which states that the change in the amount of electric charge in any volume of space is exactly equal to the amount of charge in a region and the flow of charge into and out of that region [164].

During the electrospinning process, the electrostatic repulsion between excess charges in the solution stretches the jet. This stretching also decreases the jet diameter that this leads to the law of charge conservation as the second governing equation [165].

In electrospinning process, the electric current that induced by electric field included two parts, conduction, and convection.

The conventional symbol for current is I:

$$I = I_{conduction} + I_{convection} \qquad (1.8.4)$$

Electrical conduction is the movement of electrically charged particles through a transmission medium. The movement can form an electric current in response to an electric field. The underlying mechanism for this movement depends on the material.

$$I_{conduction} = J_{cond} \times S = KE \times \pi R^2 \qquad (1.8.5)$$

$$J = \frac{I}{A(s)} \qquad (1.8.6)$$

$$I = J \times S \qquad (1.8.7)$$

Convection current is the flow of current with the absence of an electric field.

$$I_{convection} = J_{conv} \times S = 2\pi R(L) \times \sigma v \qquad (1.8.8)$$

$$J_{conv} = \sigma v \qquad (1.8.9)$$

Therefore, the total current can be calculated as:

$$\pi R^2 KE + 2\pi Rv\sigma = I \qquad (1.8.10)$$

$$\frac{\partial}{\partial t}(2\pi R\sigma) + \frac{\partial}{\partial z}(\pi R^2 KE + 2\pi Rv\sigma) = 0 \qquad (1.8.11)$$

1.10.4 MOMENTUM BALANCE

In classical mechanics, linear momentum or translational momentum is the product of the mass and velocity of an object. Like velocity, linear momentum is a vector quantity, possessing a direction as well as a magnitude:

$$P = mv \qquad (1.8.12)$$

Linear momentum is also a conserved quantity, meaning that if a closed system (one that does not exchange any matter with the outside and is not acted on by outside forces) is not affected by external forces, its total linear momentum cannot change. In classical mechanics, conservation of linear momentum is implied by Newton's laws of motion; but it also holds in special relativity (with a modified formula) and, with appropriate definitions, a (generalized) linear momentum conservation law holds in electrodynamics, quantum mechanics, quantum field theory, and general relativity[163]. For example, according to the third law, the forces between two particles are equal and opposite. If the particles are numbered 1 and 2, the second law states:

$$F_1 = \frac{dP_1}{dt} \qquad (1.8.13)$$

$$F_2 = \frac{dP_2}{dt} \qquad (1.8.14)$$

Therefore:

$$\frac{dP_1}{dt} = -\frac{dP_2}{dt} \qquad (1.8.15)$$

$$\frac{d}{dt}(P_1 + P_2) = 0 \qquad (1.8.16)$$

If the velocities of the particles are v_{11} and v_{12} before the interaction, and afterwards they are v_{21} and v_{22}, then

$$m_1 v_{11} + m_2 v_{12} = m_1 v_{21} + m_2 v_{22} \qquad (1.8.17)$$

This law holds no matter how complicated the force is between the particles. Similarly, if there are several particles, the momentum exchanged between each pair of particles adds up to zero; therefore; the total change in momentum is zero. This conservation law applies to all interactions, including collisions and separations caused by explosive forces. It can also be generalized to situations where Newton's laws do not hold, for example in the theory of relativity and in electrodynamics [153, 166]. The momentum equation for the fluid can be derived as follow:

$$\rho(\frac{dv}{dt} + v\frac{dv}{dz}) = \rho g + \frac{d}{dz}[\tau_{zz} - \tau_{rr}] + \frac{\gamma}{R^2}\cdot\frac{dr}{dz} + \frac{\sigma}{\varepsilon_0}\frac{d\sigma}{dz} + (\varepsilon - \varepsilon_0)(E\frac{dE}{dz}) + \frac{2\sigma E}{r} \quad (1.8.18)$$

But commonly, the momentum equation for electrospinning modeling is formulated by considering the forces on a short segment of the jet [153, 166].

$$\frac{d}{dz}(\pi R^2 \rho v^2) = \pi R^2 \rho g + \frac{d}{dz}\left[\pi R^2(-p + \tau_{zz})\right] + \frac{\gamma}{R}\cdot 2\pi R R' + 2\pi R(t_t^e - t_n^e R') \quad (1.8.19)$$

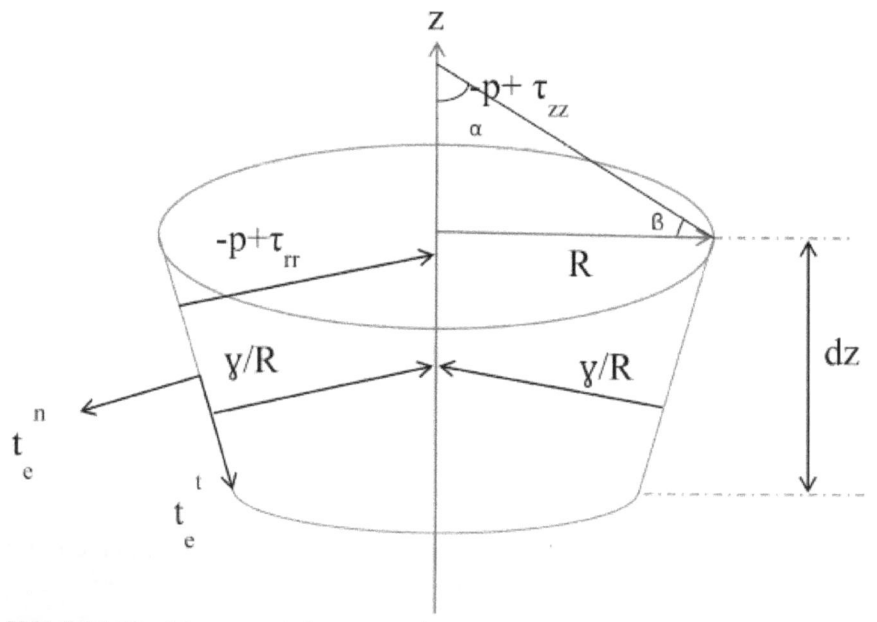

FIGURE 1.19   Momentum balance on a short section of the jet.

As shown in Figure 1.19, the element's angels could be defined as α and β. According to the mathematical relationships, it is obvious that

$$\alpha + \beta = \frac{\pi}{2} \tag{1.8.20}$$

$$\begin{aligned} \sin \alpha &= \tan \alpha \\ \cos \alpha &= 1 \end{aligned} \tag{1.8.21}$$

Due to the figure, relationships between these electrical forces are as given below:

$$t_n^e \sin \alpha \cong t_n^e \tan \alpha \cong -t_n^e \tan \beta \cong -\frac{dR}{dz} t_n^e = -R't_n^e \tag{1.8.22}$$

$$t_t^e \cos \alpha \cong t_t^e \tag{1.8.23}$$

Therefore, the effect of the electric forces in the momentum balance equation can be presented as follows:

$$2\pi RL(t_t^e - R't_n^e)dz \tag{1.8.24}$$

(Notation: In the main momentum equation, final formula is obtained by dividing into dz)

In addition, the normal electric force is defined as follows:

$$t_n^e \cong \frac{1}{2}\overline{\varepsilon}E_n^2 = \frac{1}{2}\overline{\varepsilon}(\frac{\sigma}{\overline{\varepsilon}})^2 = \frac{\sigma^2}{2\overline{\varepsilon}} \tag{1.8.25}$$

A little amount of electric forces is perished in the vicinity of the air.

$$E_n = \frac{\sigma}{\overline{\varepsilon}} \tag{1.8.26}$$

The electric force can be presented by the following equation:

$$F = \frac{\Delta We}{\Delta l} = \frac{1}{2}(\varepsilon - \overline{\varepsilon})E^2 \times \Delta S \tag{1.8.27}$$

The force per surface unit is

$$\frac{F}{\Delta S} = \frac{1}{2}(\varepsilon - \overline{\varepsilon})E^2 \qquad (1.8.28)$$

Generally the electric potential energy is obtained by using the following equation:

$$Ue = -We = -\int F.ds \qquad (1.8.29)$$

$$\Delta We = \frac{1}{2}(\varepsilon - \overline{\varepsilon})E^2 \times \Delta V = \frac{1}{2}(\varepsilon - \overline{\varepsilon})E^2 \times \Delta S.\Delta l \qquad (1.8.30)$$

Therefore, finally it could result in the following equation:

$$t_n^e = \frac{\sigma^2}{2\overline{\varepsilon}} - \frac{1}{2}(\varepsilon - \overline{\varepsilon})E^2 \qquad (1.8.31)$$

$$t_t^e = \sigma E \qquad (1.8.32)$$

1.10.5   COULOMB'S LAW

Coulomb's law is a mathematical description of the electric force between charged objects, which is formulated by the eighteenth-century French physicist Charles-Augustin de Coulomb. It is analogous to Isaac Newton's law of gravity. Both gravitational and electric forces decrease with the square of the distance between the objects, and both forces act along a line between them[167]. In Coulomb's law, the magnitude and sign of the electric force are determined by the electric charge, more than the mass of an object. Thus, a charge that is a basic property matter determines how electromagnetism affects the motion of charged targets [163].

Coulomb force is thought to be the main cause for the instability of the jet in the electrospinning process [168]. This statement is based on the Earnshaw's theorem, named after Samuel Earnshaw [169] which claims that "A charged body placed in an electric field of force cannot rest in stable equilibrium under the influence of the electric forces alone." This theorem can be notably adapted to the electrospinning process [168]. The instability of charged jet influences on jet deposition and as a consequence on nanofiber formation. Therefore, some researchers applied developed models to the analysis of mechanisms of jet deposition and alignment on various collecting devices in arbitrary electric fields [66].

The equation for the potential along the centerline of the jet can be derived from Coulomb's law. Polarized charge density is obtained

$$\rho_{p'} = -\vec{\nabla}.\vec{P}' \qquad (1.8.33)$$

where P' is polarization:

$$\vec{P}' = (\varepsilon - \overline{\varepsilon})\vec{E} \qquad (1.8.34)$$

By substituting P' in the previous equation:

$$\rho_{p'} = -(\overline{\varepsilon} - \varepsilon)\frac{dE}{dz'} \qquad (1.8.35)$$

Beneficial charge per surface unit can be calculated as follows:

$$\rho_{p'} = \frac{Q_b}{\pi R^2} \qquad (1.8.36)$$

$$Q_b = \rho_b.\pi R^2 = -(\overline{\varepsilon} - \varepsilon)\pi R^2 \frac{dE}{dz'} \qquad (1.8.37)$$

$$Q_b = -(\overline{\varepsilon} - \varepsilon)\pi\frac{d(ER^2)}{dz'} \qquad (1.8.38)$$

$$\rho_{sb} = Q_b.dz' = -(\overline{\varepsilon} - \varepsilon)\pi\frac{d}{dz'}(ER^2)dz' \qquad (1.8.39)$$

The main equation of Coulomb's law:

$$F = \frac{1}{4\pi\varepsilon_0}\frac{qq_0}{r^2} \qquad (1.8.40)$$

The electric field is:

$$E = \frac{1}{4\pi\varepsilon_0}\frac{q}{r^2}$$

(1.8.41)

The electric potential can be measured:

$$\Delta V = -\int E.dL$$

(1.8.42)

$$V = \frac{1}{4\pi\varepsilon_0}\frac{Q_b}{r}$$

(1.8.43)

According to the beneficial charge equation, the electric potential could be rewritten as follows:

$$\Delta V = Q(z) - Q_\infty(z) = \frac{1}{4\pi\bar\varepsilon}\int\frac{(q-Q_b)}{r}dz'$$

(1.8.44)

$$Q(z) = Q_\infty(z) + \frac{1}{4\pi\bar\varepsilon}\int\frac{q}{r}dz' - \frac{1}{4\pi\bar\varepsilon}\int\frac{Q_b}{r}dz'$$

(1.8.45)

$$Q_b = -(\bar\varepsilon - \varepsilon)\pi\frac{d(ER^2)}{dz'}$$

(1.8.46)

The surface charge density's equation is

$$q = \sigma.2\pi RL$$

(1.8.47)

$$r^2 = R^2 + (z - z')^2$$

(1.8.48)

$$r = \sqrt{R^2 + (z - z')^2}$$

(1.8.49)

The final equation that obtained by substituting the mentioned equations is

$$Q(z) = Q_\infty(z) + \frac{1}{4\pi\bar{\varepsilon}} \int \frac{\sigma.2\pi R}{\sqrt{(z-z')^2 + R^2}} dz' - \frac{1}{4\pi\bar{\varepsilon}} \int \frac{(\bar{\varepsilon} - \varepsilon)\pi}{\sqrt{(z-z')^2 + R^2}} \frac{d(ER^2)}{dz'} \quad (1.8.50)$$

It is assumed that β is defined as follows:

$$\beta = \frac{\varepsilon}{\bar{\varepsilon}} - 1 = -\frac{(\bar{\varepsilon} - \varepsilon)}{\bar{\varepsilon}} \quad (1.8.51)$$

Therefore, the potential equation becomes

$$Q(z) = Q_\infty(z) + \frac{1}{2\bar{\varepsilon}} \int \frac{\sigma.R}{\sqrt{(z-z')^2 + R^2}} dz' - \frac{\beta}{4} \int \frac{1}{\sqrt{(z-z')^2 + R^2}} \frac{d(ER^2)}{dz'} \quad (1.8.52)$$

The asymptotic approximation of χ is used to evaluate the integrals mentioned earlier:

$$\chi = \left(-z + \xi + \sqrt{z^2 - 2z\xi + \xi^2 + R^2} \right) \quad (1.8.53)$$

where χ is "aspect ratio" of the jet (L = length, R_0 = initial radius)

This leads to the final relation to the axial electric field:

$$E(z) = E_\infty(z) - \ln \chi \left(\frac{1}{\bar{\varepsilon}} \frac{d(\sigma R)}{dz} - \frac{\beta}{2} \frac{d^2(ER^2)}{dz^2} \right) \quad (1.8.54)$$

1.10.6 FORCES CONSERVATION

There exists a force, as a result of charge build-up, acting on the droplet coming out of the syringe needle pointing toward the collecting plate which can be either grounded or oppositely charged. Further, similar charges within the droplet promote jet initiation due to their repulsive forces. Nevertheless, surface tension and other hydrostatic forces inhibit the jet initiation because the total energy of a droplet is lower than that of a thin jet of equal volume upon consideration of surface energy. When the forces that aid jet initiation (such as electric field and Columbic) overcome the opposing forces (such as surface tension and gravitational), the droplet accelerates toward the collecting plate. This forms a jet of very small diameter. Other than initiating jet flow, the electric field and Columbic forces tend to stretch the jet, thereby contributing toward the thinning effect of the resulting nanofibers.

In the flow path modeling, we recall the Newton's Second Law of motion:

$$m\frac{d^2P}{dt^2} = \Sigma f \qquad (1.8.55)$$

where, m (equivalent mass) and the various forces are summed as follows:

$$\Sigma f = f_C + f_E + f_V + f_S + f_A + f_G + ... \qquad (1.8.56)$$

In which subscripts C, E, V, S, A and G correspond to the Columbic, electric field, viscoelastic, surface tension, air drag, and gravitational forces, respectively. A description of each of these forces based on the literature [5] is summarized in Table 1.1 where

V_0 = applied voltage
h = distance from pendent drop to ground collector
σ_V = viscoelastic stress
v = kinematic viscosity

TABLE 1.1   Description of itemized forces or terms related to them

Forces	Equations
Columbic	$f_C = \dfrac{q^2}{l^2}$
Electric field	$f_E = -\dfrac{qV_0}{h}$
Viscoelastic	$f_V = \dfrac{d\sigma_V}{dt} = \dfrac{G}{l}\dfrac{dl}{dt} - \dfrac{G}{\eta}\sigma_V$
Surface tension	$f_S = \dfrac{\alpha\pi R^2 k}{\sqrt{x_i^2 + y_i^2}}\left[i|x|Sin(x) + i|y|Sin(y)\right]$

Air drag	$f_A = 0.65\pi R \rho_{air} v^2 \left(\dfrac{2vR}{v_{air}}\right)^{-0.81}$
Gravitational	$f_G = \rho g \pi R^2$

1.10.7 CONSTITUTIVE EQUATIONS

In modern condensed matter physics, the constitutive equation plays a major role. In physics and engineering, a constitutive equation or relation is a relation between two physical quantities that is specific to a material or substance, and approximates the response of that material to external stimulus, usually as applied fields or forces [170]. There are a sort of mechanical equation of state, which describe how the material is constituted mechanically. With these constitutive relations, the vital role of the material is reasserted [171]. There are two groups of constitutive equations: Linear and nonlinear constitutive equations [172]. These equations are combined with other governing physical laws to solve problems; for example, in fluid mechanics the flow of a fluid in a pipe, in solid-state physics the response of a crystal to an electric field, or in structural analysis, the connection between applied stresses or forces to strains or deformations [170].

The first constitutive equation (constitutive law) was developed by Robert Hooke and is known as Hooke's law. It deals with the case of linear elastic materials. Following this discovery, this type of equation, often called a "stress-strain relation" in this example, but also called a "constitutive assumption" or an "equation of state" was commonly used [173]. Walter Noll advanced the use of constitutive equations, clarifying their classification and the role of invariance requirements, constraints, and definitions of terms such as "material," "isotropic," "aeolotropic," and so on. The class of "constitutive relations" of the form stress rate = f (velocity gradient, stress, density) was the subject of Walter Noll's dissertation in 1954 under Clifford Truesdell [170]. There are several kinds of constitutive equations that are applied commonly in electrospinning. Some of these applicable equations are discussed in the following.

1.10.7.1   OSTWALD–DE WAELE POWER LAW

Rheological behavior of many polymer fluids can be described by power law constitutive equations [172]. The equations that describe the dynamics in electrospinning constitute, at a minimum, those describing the conservation of mass, momentum and charge, and the electric field equation. In addition, a constitutive equation for the fluid behavior is also required [76]. A power law fluid, or the Ostwald–de Waele —relationship, is a type of generalized Newtonian fluid for which the shear stress, τ, is given by

$$\tau = K' \left(\frac{\partial v}{\partial y} \right)^{m} \tag{1.8.57}$$

where $\partial v/\partial y$ is the shear rate or the velocity gradient perpendicular to the plane of shear. The power law is only a good description of fluid behavior across the range of shear rates to which the coefficients are fitted. There are a number of other models that better describe the entire flow behavior of shear-dependent fluids, but they do so at the expense of simplicity; therefore, the power law is still used to describe fluid behavior, permit mathematical predictions, and correlate experimental data [166, 174].

Nonlinear rheological constitutive equations applicable for polymer fluids (Ostwald–de Waele power law) were applied to the electrospinning process by Spivak and Dzenis [77, 150, 175].

$$\hat{\tau}^{c} = \mu \left[tr \left(\dot{\hat{\gamma}}^{2} \right) \right]^{(m-1)/2} \dot{\hat{\gamma}} \tag{1.8.58}$$

$$\mu = K \left(\frac{\partial v}{\partial y} \right)^{m-1} \tag{1.8.59}$$

Viscous Newtonian fluids are described by a special case of equation above with the flow index $m = 1$. Pseudoplastic (shear thinning) fluids are described by flow indices $0 \leq m \leq 1$. Dilatant (shear thickening) fluids are described by the flow indices $m > 1$ [150].

1.10.7.2   GIESEKUS EQUATION

In 1966, Giesekus established the concept of anisotropic forces and motions in polymer kinetic theory. With particular choices for the tensors describing the anisotropy, one can obtain Giesekus constitutive equation from elastic dumbbell kinetic theory [176–177]. The Giesekus equation is known to predict, both qualitatively and quantitatively, material functions for steady and nonsteady shear and elongational flows.

However, the equation sustains two drawbacks: it predicts that the viscosity is inversely proportional to the shear rate in the limit of infinite shear rate and it is unable to predict any decrease in the elongational viscosity with increasing elongation rates in uniaxial elongational flow. The first one is not serious because of the retardation time that is included in the constitutive equation, but the second one is more critical because the elongational viscosity of some polymers decreases with increasing of elongation rate [178–179].

In the main Giesekus equation, the tensor of excess stresses depending on the motion of polymer units relative to their surroundings was connected to a sequence of tensors characterizing the configurational state of the different kinds of network structures present in the concentrated solution or melt. The respective set of constitutive equations indicates [180–181]:

$$S_k + \eta \frac{\partial C_k}{\partial t} = 0 \qquad (1.8.60)$$

The equation below indicates the upper-convected time derivative (Oldroyd derivative):

$$\frac{\partial C_k}{\partial t} = \frac{DC_k}{Dt} - \left[C_k \nabla v + (\nabla v)^T C_k \right] \qquad (1.8.61)$$

(Note: The upper convective derivative is the rate of change of any tensor property of a small parcel of fluid that is written in the coordinate system rotating and stretching with the fluid.)

C_k also can be measured as follows:

$$C_k = 1 + 2E_k \qquad (1.8.62)$$

According to the concept of "recoverable strain" S_k may be understood as a function of E_k and vice versa. If linear relations corresponding to Hooke's law are adopted.

$$S_k = 2\mu_k E_k \qquad (1.8.63)$$

Therefore,

$$S_k = \mu_k (C_k - 1) \tag{1.8.64}$$

Equation (1.8.60) becomes thus:

$$S_k + \lambda_k \frac{\partial S_k}{\partial t} = 2\eta D \tag{1.8.65}$$

$$\lambda_k = \frac{\eta}{\mu_k} \tag{1.8.66}$$

As a second step in order to rid the model of the shortcomings is the scalar mobility constants B_k, which are contained in the constants η. This mobility constant can be represented as follows:

$$\tfrac{1}{2}(\beta_k S_k + S_k \beta_k) + \breve{\eta} \frac{\partial C_k}{\partial t} = 0 \tag{1.8.67}$$

The two parts of equation (1.8.67) reduces to the single constitutive equation:

$$\beta_k + \breve{\eta} \frac{\partial C_k}{\partial t} = 0 \tag{1.8.68}$$

The excess tension tensor in the deformed network structure where the well-known constitutive equation of a so-called Neo–Hookean material is proposed [180, 182]:

Neo–Hookean equation

$$S_k = 2\mu_k E_k = \mu_k (C_k - 1) \tag{1.8.69}$$

$$\mu_k = NKT$$
$$\beta_k = 1 + \alpha(C_k - 1) = (1 - \alpha) + \alpha C_k \tag{1.8.70}$$

where K is Boltzmann's constant.

By substitution Eqs (1.8.69) and (1.8.70) in the Eq. (1.8.64), it can be obtained where the condition $0 \leq \alpha \leq 1$ must be fulfilled, the limiting case $\alpha = 0$ corresponds to an isotropic mobility [183].

$$0 \leq \alpha \leq 1 \quad [1 + \alpha(C_k - 1)](C_k - 1) + \lambda_k \frac{\partial C_k}{\partial t} = 0 \qquad (1.8.71)$$

$$\alpha = 1 \qquad C_k(C_k - 1) + \lambda_k \frac{\partial C_k}{\partial t} = 0 \qquad (1.8.72)$$

$$0 \leq \alpha \leq 1 \qquad C_k = \frac{S_k}{\mu_k} + 1 \qquad (1.8.73)$$

By substituting equations above in Eq. (1.8.64), we obtain

$$\left[1 + \frac{\alpha S_k}{\mu_k}\right] \frac{S_k}{\mu_k} + \lambda_k \frac{\partial C_k}{\partial t} = 0 \qquad (1.8.74)$$

$$\frac{S_k}{\mu_k} + \frac{\alpha S_k^2}{\mu_k^2} + \lambda_k \frac{\partial (S_k/\mu_k + 1)}{\partial t} = 0 \qquad (1.8.75)$$

$$\frac{S_k}{\mu_k} + \frac{\alpha S_k^2}{\mu_k^2} + \frac{\lambda_k}{\mu_k} \frac{\partial S_k}{\partial t} = 0 \qquad (1.8.76)$$

$$S_k + \frac{\alpha S_k^2}{\mu_k} + \lambda_k \frac{\partial S_k}{\partial t} = 0 \qquad (1.8.77)$$

D means the rate of strain tensor of the material continuum [180].

$$D = \frac{1}{2}\left[\nabla \upsilon + (\nabla \upsilon)^T\right] \qquad (1.8.78)$$

The equation of the upper convected time derivative for all fluid properties can be calculated as follows:

$$\frac{\partial \otimes}{\partial t} = \frac{D \otimes}{Dt} - \left[\otimes . \nabla \upsilon + (\nabla \upsilon)^T . \otimes\right] \qquad (1.8.79)$$

$$\frac{D\otimes}{Dt} = \frac{\partial\otimes}{\partial t} + \left[(v.\nabla).\otimes\right] \qquad (1.8.80)$$

By replacing S_k instead of the symbol:

$$\lambda_k \frac{\partial S_k}{\partial t} = \lambda_k \frac{DS_k}{Dt} - \lambda_k \left[S_k\nabla v + (\nabla v)^T S_k\right] = \lambda_k \frac{DS_k}{Dt} - \lambda_k(v.\nabla)S_k \qquad (1.8.81)$$

By simplification the equation above, we obtain

$$S_k + \frac{\alpha S_k^2}{\mu_k} + \lambda_k \frac{DS_k}{Dt} = \lambda_k(v.\nabla)S_k \qquad (1.8.82)$$

$$S_k = 2\mu_k E_k \qquad (1.8.83)$$

The assumption of $E_k = 1$ would lead to the next equation:

$$S_k + \frac{\alpha\lambda_k S_k^2}{\eta} + \lambda_k \frac{DS_k}{Dt} = \frac{\eta}{\mu_k}(2\mu_k)D = 2\eta D = \eta\left[\nabla v + (\nabla v)^T\right] \qquad (1.8.84)$$

In electrospinning modeling articles τ is used commonly instead of S_k [154, 159, 161].

$$S_k \leftrightarrow \tau$$

$$\tau + \frac{\alpha\lambda_k \tau^2}{\eta} + \lambda_k \tau_{(1)} = \eta\left[\nabla v + (\nabla v)^T\right] \qquad (1.8.85)$$

1.10.7.3   MAXWELL EQUATION

Maxwell's equations are a set of partial differential equations that, together with the Lorentz force law, form the foundation of classical electrodynamics, classical optics, and electric circuits. These fields are the bases of modern electrical and communications technologies. Maxwell's equations describe how electric and magnetic fields are generated and altered by each other and by charges and currents; they are named after the Scottish physicist and mathematician James Clerk Maxwell who published an early form of those equations between 1861 and 1862 [184–185]. It will be discussed further in detail.

1.10.8 MICROSCOPIC MODELS

One of the aims of computer simulation is to reproduce experiment to elucidate the invisible microscopic details and further explain the experiments. Physical phenomena occurring in complex materials cannot be encapsulated within a single numerical paradigm. In fact, they should be described within hierarchical, multilevel numerical models in which each submodel is responsible for different spatial-temporal behavior and passes out the averaged parameters of the model, which is next in the hierarchy. The understanding of the nonequilibrium properties of complex fluids such as the viscoelastic behavior of polymeric liquids, the rheological properties of ferrofluids and liquid crystals subjected to magnetic fields, based on the architecture of their molecular constituents is useful to get a comprehensive view of the process. The analysis of simple physical particle models for complex fluids has developed from the molecular computation of basic systems (atoms, rigid molecules) to the simulation of macromolecular "complex" system with a large number of internal degrees of freedom exposed to external forces [186–187].

The most widely used simulation methods for molecular systems are Monte Carlo, Brownian dynamics, and molecular dynamics. The microscopic approach represents the microstructural features of material by means of a large number of micromechanical elements (beads, platelet, rods) obeying stochastic differential equations. The evolution equations of the microelements arise from a balance of momentum at the elementary level. The Monte Carlo method is a stochastic strategy that relies on probabilities. The Monte Carlo sampling technique generates large numbers of configurations or microstates of equilibrated systems by stepping from one microstate to the next in a particular statistical ensemble. Random changes are made to the positions of the species present, together with their orientations and conformations where appropriate. Brownian dynamics are an efficient approach for simulations of large polymer molecules or colloidal particles in a small molecule solvent. Molecular dynamics is the most detailed molecular simulation method which computes the motions of individual molecules. Molecular dynamics efficiently evaluates different configurational properties and dynamic quantities which cannot generally be obtained by Monte Carlo [188–189].

The first computer simulation of liquids was carried out in 1953. The model was an idealized two-dimensional representation of molecules as rigid disks. For macromolecular systems, the coarse-grained approach is widely used as the modeling process is simplified, hence becomes more efficient, and the characteristic topological features of the molecule can still be maintained. The level of detail for a coarse-grained model varies in different cases. The whole molecule can be represented by a single particle in a simulation and interactions between particles incorporate average properties of the whole molecule. With this approach, the number of degrees of freedom is greatly reduced [190].

On the contrary, a segment of a polymer molecule can also be represented by a particle (bead). The first coarse-grained model, called the "dumbbell" model, was introduced in the 1930s. Molecules are treated as a pair of beads interacting via a harmonic potential. However by using this model, it is possible to perform kinetic theory derivations and calculations for nonlinear rheological properties and solve some flow problems. The analytical results for the dumbbell models (Figure 1.20) can also be used to check computer simulation procedures in molecular dynamics and Brownian dynamics [191–192].

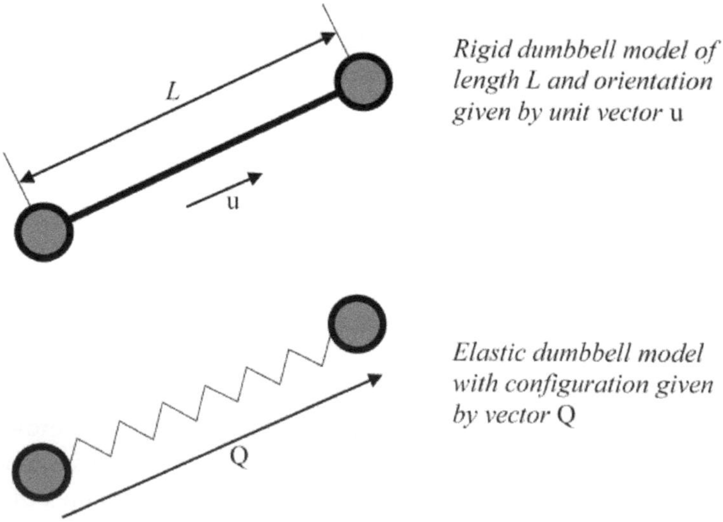

Rigid dumbbell model of length L and orientation given by unit vector u

Elastic dumbbell model with configuration given by vector Q

FIGURE 1.20   The first coarse-grained models—the rigid and elastic dumbbell models.

The bead-rod and bead-spring model (Figure 1.21) were introduced to model chainlike macromolecules. Beads in the bead-rod model do not represent the atoms of the polymer chain backbone, but some portion of the chain, normally 10 to 20 monomer units. These beads are connected by rigid and massless rods. While in the bead-spring model, a portion of the chain containing several hundreds of backbone atoms are replaced by a "spring," and the masses of the atoms are concentrated on the mass of beads [193].

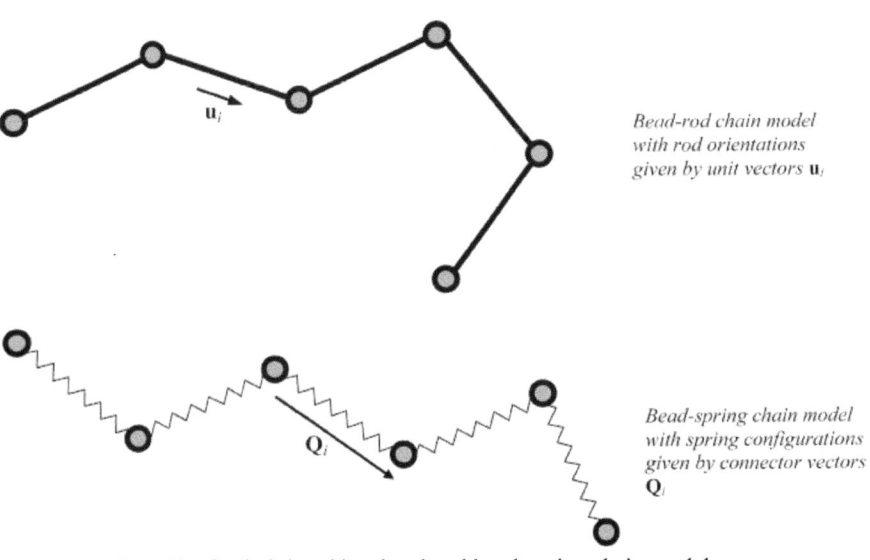

Bead-rod chain model with rod orientations given by unit vectors $\mathbf{u}_i$

Bead-spring chain model with spring configurations given by connector vectors $\mathbf{Q}_i$

FIGURE 1.21   The freely jointed bead-rod and bead-spring chain models.

If the springs are taken to be Hookean springs, the bead-spring chain is referred to as a Rouse chain or a Rouse–Zimm chain. This approach has been applied widely as it has a large number of internal degrees of freedom and exhibits orientability and stretchability. However the disadvantage of this model is that it does not have a constant contour length and can be stretched out to any length. Therefore, in many cases finitely extensible springs with two more parameters, the spring constant and the maximum extensibility of an individual spring, can be included so the contour length of the chain model cannot exceed a certain limit [194–195].

The understanding of the nonequilibrium properties of complex fluids such as the viscoelastic behavior of polymeric liquids, the rheological properties of ferrofluids and liquid crystals subjected to magnetic fields, based on the architecture of their molecular constituents [186].

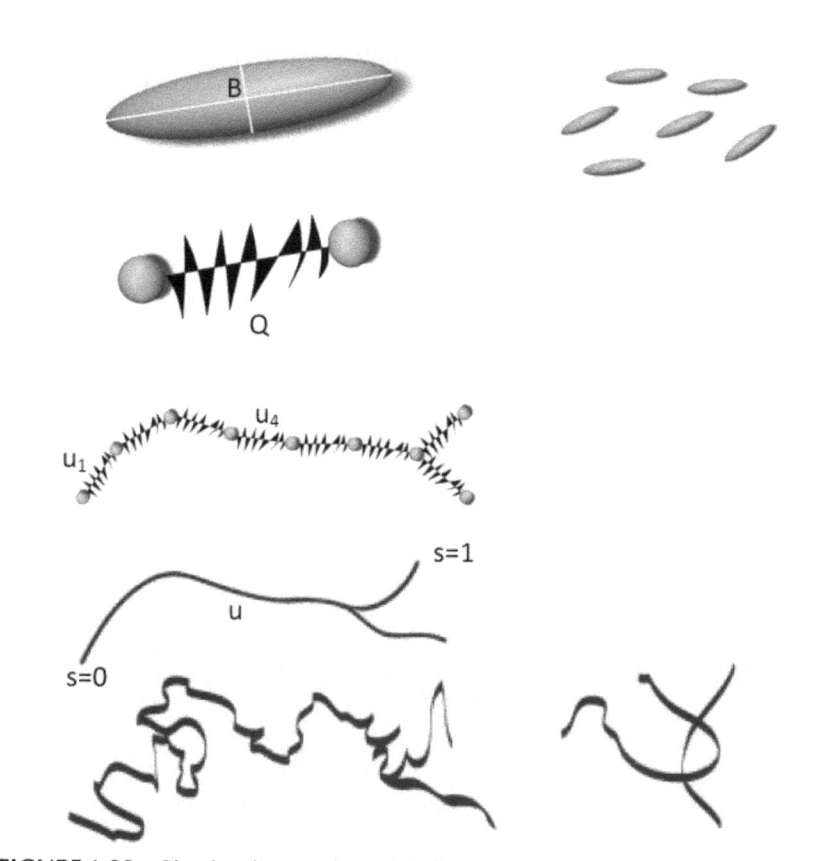

FIGURE 1.22   Simple microscopic models for complex fluids by using dumbbell model.

Dumbbell models are very crude representations of polymer molecules. Too crude to be of much interest to a polymer chemist, since it in no way accounts for the details of the molecular architecture. It certainly does not have enough internal degrees of freedom to describe the very rapid motions that contribute, for example, to the complex viscosity at high frequencies. On the contrary, the elastic dumbbell model is orientable and stretchable, and these two properties are essential for the qualitative description of steady-state rheological properties and those involving slow changes with time. For dumbbell models, one can go through the entire program of endeavor—from molecular model for fluid dynamics—

for illustrative purposes, in order to point the way toward the task that has ultimately to be performed for more realistic models. According to the researches, dumbbell models must, to some extent then, be regarded as mechanical playthings, somewhat disconnected from the real world of polymers (Figure 1.22). However, when used intelligently , they can be useful pedagogically and very helpful in developing a qualitative understanding of rheological phenomena [186, 196].

The simplest model of flexible macromolecules in a dilute solution is the elastic dumbbell (or bead-spring) model. This has been widely used for purely mechanical theories of the stress in electrospinning modeling [197].

A Maxwell constitutive equation was first applied by Reneker et al. in 2000. Consider an electrified liquid jet in an electric field parallel to its axis. They modeled a segment of the jet by a viscoelastic dumbbell (Figure 1.23). They used a Gaussian electrostatic system of units. According to this model each particle in the electric field exerts repulsive force on another particle [66].

He had three main assumptions [66, 198]:

1. The background electric field created by the generator is considered static.
2. The fiber is a perfect insulator.
3. The polymer solution is a viscoelastic medium with constant elastic modulus, viscosity, and surface tension.

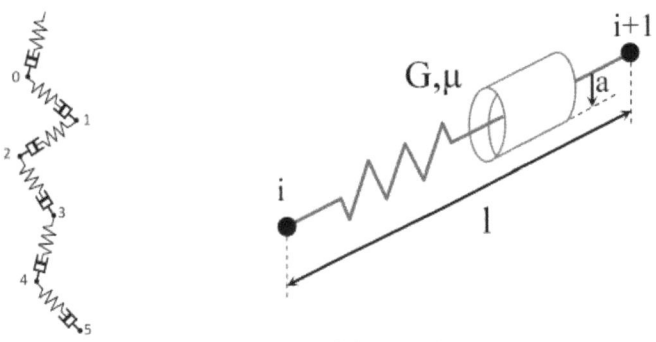

FIGURE 1.23   A schematic of one section of the model.

The researcher considered the governing equations for each bead as follows [198]:

$$\frac{d}{dt}\left(\pi a^2 l\right) = 0 \tag{1.8.86}$$

Therefore, the stress between these particles can be measured by using the following equation [66]:

$$\frac{d\sigma}{dt} = G\frac{dl}{ldt} - \frac{G}{\eta}\sigma \tag{1.8.87}$$

The stress can be calculated by a Maxwell viscoelastic constitutive equation [199]:

$$\dot{\tau} = G\left(\varepsilon' - \frac{\tau}{\eta}\right) \tag{1.8.88}$$

where ε' is the Lagrangian axial strain [199]:

$$\varepsilon' \equiv \frac{\partial \dot{x}}{\partial \xi}.\hat{t}. \tag{1.8.89}$$

Equation of motion for beads can be written as follows[200]:

$$\text{mass} \times \text{acceleration} = \text{viscous drag} + \text{Brownian motion force} \tag{8.90}$$
$$+ \text{ force of one bead on another through the connector}$$

The momentum balance for a bead is [198]:

$$m\frac{dv}{dt} = \underbrace{-\frac{q^2}{l^2}}_{Coulomb\ \ forces} - \underbrace{qE}_{Electric\ \ force} + \underbrace{\pi a^2 \sigma}_{Mechanical\ \ forces} \tag{1.8.91}$$

So the momentum conservation for model charges can be calculated as [201] follows:

$$m_i\frac{dv_i}{dt} = \underbrace{q_i\sum_{i \neq j} q_j K \frac{r_i - r_j}{\left|r_i - r_j\right|^3}}_{Coulomb\ \ forces} + \underbrace{q_i E}_{Electric\ \ force} + \underbrace{\pi a_{i,i+1}^2 \sigma_{i,i+1}\frac{r_{i+1} - r_i}{\left|r_{i+1} - r_i\right|} - \pi a_{i-1,i}^2 \sigma_{i-1,i}\frac{r_i - r_{i-1}}{\left|r_i - r_{i-1}\right|}}_{Mechanical\ \ forces} \tag{1.8.92}$$

Boundary condition assumptions: A small initial perturbation is added to the position of the first bead, the background electric field is axial and uniform and the first bead is described by a stationary equation. For solving these equations, some dimensionless parameters are defined then by simplifying equations, the equations are solved by using boundary conditions [198, 201].

Now, an example for using this model for the polymer structure is mentioned. For a dumbbell consists of two that are connected with a nonlinear spring (Figure 1.24), the spring force law is given by the following equation[96]:

$$F = -\frac{HQ}{1 - Q^2/Q_0^2} \tag{1.8.93}$$

Now if we considered the model for the polymer matrix such as carbon nanotube, the rheological behavior can be obtained as follows [96, 201]:

$$\tau_{ij} = \tau_p + \tau_s \tag{1.8.94}$$

$$\tau_p = \underbrace{n_a \langle Q_a F_a \rangle}_{aggregated\ dumbbells} + \underbrace{n_f \langle Q_f F_f \rangle}_{free\ dumbbells} - nkT\delta_{ij} \tag{1.8.95}$$

$$\tau_s = \eta\dot{\gamma} \tag{1.8.96}$$

$$\lambda\langle Q.Q \rangle^{\nabla} = \delta_{ij} - \frac{c\langle Q.Q \rangle}{1 - tr\langle Q.Q \rangle/b_{max}} \tag{1.8.97}$$

The polymeric stress can be obtained from the following relation [96]:

$$\frac{\hat{\tau}_{ij}}{n_d kT} = \delta_{ij} - \frac{c\langle Q.Q \rangle}{1 - tr\langle Q.Q \rangle/b_{max}} \tag{1.8.98}$$

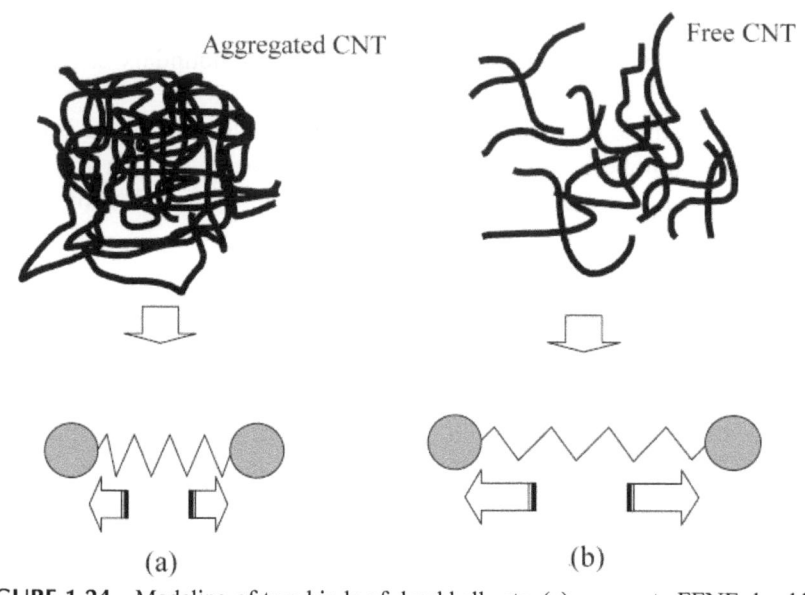

FIGURE 1.24   Modeling of two kinds of dumbbell sets, (a) aggregate FENE dumbbell which has lower mobility and (b) free FENE dumbbell which has higher mobility.

1.10.9   SCALING

The physical aspect of a phenomenon can use the language of differential equation that represents the structure of the system by selecting the variables that characterize the state of it and certain mathematical constraint on the values of those variables can take on. These equations can predict the behavior of the system over a quantity such as time. For an instance, a set of continuous functions of time that describe the way the variables of the system developed over time starting from a given initial state [208]. In general, the renormalization group theory, scaling, and fractal geometry are applied to the understanding of the complex phenomena in physics, economics, and medicine [203-209].

In more recent times, in statistical mechanics, the expression "scaling laws" has referred to the homogeneity of form of the thermodynamic and correlation functions near critical points, and to the resulting relations among the exponents that occur in those functions. From the viewpoint of scaling, electrospinning modeling can be studied in two ways: allometric and dimensionless analysis. Scaling and dimensional analysis actually started with Newton, and allometry exists everywhere in our daily life and scientific activity[209–210].

1.10.9.1   ALLOMETRIC SCALING

Electrospinning applies electrically generated motion to spin fibers. Therefore, it is difficult to predict the size of the produced fibers, which depends on the applied voltage in principal. Therefore, the relationship between the radius of the jet and the axial distance from the nozzle is always the subject of investigation [211–212]. It can be described as an allometric equation by using the values of the scaling exponent for the initial steady, instability, and terminal stages [213].

The relationship between r and z can be expressed as an allometric equation of the following form:

$$r \approx z^b \qquad (1.8.99)$$

When the power exponent, $b = 1$, the relationship is isometric and when $b \neq 1$ the relationship is allometric [211, 214]. In another view, $b = -1/2$ is considered for the straight jet, $b = -1/4$ for instability jet, and $b = 0$ for final stage [172, 212].

Due to high electrical force acting on the jet, it can be illustrated [211]:

$$\frac{d}{dz}\left(\frac{v^2}{2}\right) = \frac{2\sigma E}{\rho r} \qquad (1.8.100)$$

Equations of mass and charge conservations applied here as mentioned earlier [211, 214–215]

From the above equations, it can be noted that [161, 211]

$$r \approx z^b, \sigma \approx r, E \approx r^{-2}, \frac{dv^2}{dz} \approx r^{-2} \qquad (1.8.101)$$

Therefore, it is obtained for the initial part of jet, $r \approx z^{-1/4}$ for the instable stage and for the final stage.

The charged jet can be considered as a 1D flow as mentioned. If the conservation equations modified, they would change as follows [211]:

$$2\pi r \sigma^{\alpha} v + K \pi r^2 E = I \qquad (1.8.102)$$

$$r \approx z^{-\alpha/(\alpha+1)} \qquad (1.8.103)$$

where α is a surface charge parameter; the value of α depends on the surface charge in the jet. When $\alpha = 0$ no charge in jet surface, and in $\alpha = 1$ use for full surface charge.

Allometric scaling equations are more widely investigated by different researchers. Some of the most important allometric relationships for electrospinning are presented in Table 1.2.

TABLE 1.2   Investigated scaling laws applied in electrospinning model

Parameters	Equation	Ref.
Conductance and polymer concentration	$g \approx c^{\beta}$	[161]
Fiber diameters and the solution viscosity	$d \approx \eta^{\alpha}$	[212]
Mechanical strength and threshold voltage	$\bar{\sigma} \approx E_{threshold}^{-\alpha}$	[216]
Threshold voltage and the solution viscosity	$E_{threshold} \approx \eta^{1/4}$	[216]
Viscosity and the oscillating frequency	$\eta \approx \omega^{-0.4}$	[216]
Volume flow rate and the current	$I \approx Q^{b}$	[215]
Current and the fiber radius	$I \approx r^{2}$	[217]
Surface charge density and the fiber radius	$\sigma \approx r^{3}$	[217]
Induction surface current and the fiber radius	$\phi \approx r^{2}$	[217]
Fiber radius and AC frequency	$r \approx \Omega^{1/4}$	[172]

Note: β, α, and b = scaling exponent

1.10.9.2 DIMENSIONLESS ANALYSIS

One of the simplest, yet most powerful, tools in the physics is dimensional analysis in which there are two kinds of quantities: dimensionless and dimensional.

In physics and all science, dimensional analysis is the analysis of the relationships between different physical quantities by identifying their dimensions. The dimension of any physical quantity is the combination of the basic physical dimensions that compose it, although the definitions of basic physical dimensions may vary. Some fundamental physical dimensions, based on the SI system of units, are length, mass, time, and electric charge. (The SI unit of electric charge is, however, defined in terms of units of length, mass and time, and, for example, the time unit and the length unit are not independent but can be linked by the speed of light c.) Other physical quantities can be expressed in terms of these fundamental physical dimensions. Dimensional analysis is based on the fact that a physical law must be independent of the units used to measure the physical variables. A straightforward practical consequence is that any meaningful equation (and any inequality and inequation) must have the same dimensions on the left and right sides. Dimensional analysis is routinely used as a check on the plausibility of derived equations and computations. It is also used to categorize types of physical quantities and units based on their relationship to or dependence on other units.

Dimensionless quantities that are without associated physical dimensions are widely used in mathematics, physics, engineering, economics, and in everyday life (such as in counting). Numerous well-known quantities, such as π, e, and φ, are dimensionless. They are "pure" numbers, and as such always have a dimension of 1 [218–219].

Dimensionless quantities are often defined as products or ratios of quantities that are not dimensionless, but whose dimensions cancel out when their powers are multiplied [220].

The basic principle of dimensional analysis was known to Isaac Newton (1686) who referred to it as the "Great Principle of Similitude." James Clerk Maxwell played a major role in establishing modern use of dimensional analysis by distinguishing mass, length, and time as fundamental units, while referring to other units as derived. The nineteenth-century French mathematician Joseph Fourier made important contributions based on the idea that physical laws like $F = ma$ should be independent of the units used to measure the physical variables. This led to the conclusion that meaningful laws must be homogeneous equations in their various units of measurement, a result that was eventually formalized in the Buckingham π theorem. This theorem describes how every physically meaningful equation involving n variables can be equivalently rewritten as an equation of $n-m$ dimensionless parameters, where m is the rank of the dimensional matrix. Further, and most importantly, it provides a method for computing these dimensionless parameters from the given variables.

A dimensional equation can have the dimensions reduced or eliminated through nondimensionalization, which begins with dimensional analysis, and involves scaling quantities by characteristic units of a system or natural units of nature. This gives insight into the fundamental properties of the system, as illustrated in the examples below:

In nondimensional scaling, there are two key steps:
1. Identifying a set of physically relevant dimensionless groups
2. Determining the scaling exponent for each one

Dimensional analysis will help you with step (a), but it cannot be applicable possibly for step (b).

A good approach to systematically getting to grips with such problems is through the tools of dimensional analysis (Bridgman, 1963). The dominant balance of forces controlling the dynamics of any process depends on the relative magnitudes of each underlying physical effect entering the set of governing equations [221]. Now, the most general characteristics parameters that are used in dimensionless analysis in electrospinning are introduced in Table 1.3.

TABLE 1.3   Characteristics parameters employed and their definitions

Parameter	Definition
Length	R_0
Velocity	$v_0 = \dfrac{Q}{\pi R_0^2 K}$
Electric field	$E_0 = \dfrac{I}{\pi R_0^2 K}$
Surface charge density	$\sigma_0 = \bar{\varepsilon} E_0$
Viscous stress	$\tau_0 = \dfrac{\eta_0 v_0}{R_0}$

For achievement of a simplified form of equations and reduction of a number of unknown variables, the parameters should be subdivided into characteristic scales to become dimensionless. Electrospinning dimensionless groups are shown in Table 1.4 [222].

TABLE 1.4   Dimensionless groups employed and their definitions

Name	Definition	Field of application
Froude number	$Fr = \dfrac{v_0^2}{gR_0}$	The ratio of inertial to gravitational forces
Reynolds number	$Re = \dfrac{\rho v_0 R_0}{\eta_0}$	The ratio of the inertia forces of the viscous forces
Weber number	$We = \dfrac{\rho v_0^2 R_0}{\gamma}$	The ratio of the surface tension forces to the inertia forces
Deborah number	$De = \dfrac{\lambda v_0}{R_0}$	The ratio of the fluid relaxation time to the instability growth time
Electric Peclet number	$Pe = \dfrac{2\bar{\varepsilon} v_0}{KR_0}$	The ratio of the characteristic time for flow to that for electrical conduction
Euler number	$Eu = \dfrac{\varepsilon_0 E^2}{\rho v_0^2}$	The ratio of electrostatic forces to inertia forces
Capillary number	$Ca = \dfrac{\eta v_0}{\gamma}$	The ratio of inertia forces of viscous forces
Ohnesorge number	$oh = \dfrac{\eta}{\left(\rho \gamma R_0\right)^{1/2}}$	The ratio of viscous force to surface force
Viscosity ratio	$r_\eta = \dfrac{\eta_p}{\eta_0}$	The Ratio of the Polymer Viscosity to Total Viscosity
Aspect ratio	$\chi = \dfrac{L}{R_0}$	The ratio of the length of the primary radius of jet

TABLE 1.4   *(Continued)*

Name	Definition	Field of application
Electrostatic force parameter	$\varepsilon = \dfrac{\overline{\varepsilon} E_0^2}{\rho v_0^2}$	The relative importance of the electrostatic and hydrodynamic forces
Dielectric constant ratio	$\beta = \dfrac{\varepsilon}{\overline{\varepsilon}} - 1$	The ratio of the field without the dielectric to the net field with the dielectric

     The governing and constitutive equations can be transformed into a dimensionless form using the dimensionless parameters and groups.

1.10.10   SOME OF ELECTROSPINNING MODELS

The most important mathematical models for electrospinning process are classified in the Table 1.5 according to the year, advantages, and disadvantages of the models.

TABLE 1.5   The most important mathematical models for electrospinning

Researchers	Model	Year	Ref.
Taylor, G. I. Melcher, J. R.	Leaky dielectric model ✓ Dielectric fluid ✓ Bulk charge in the fluid jet considered to be zero ✓ Only axial motion ✓ Steady-state part of jet	1969	[223]
Ramos	Slender body ✓ Incompressible and axisymmetric and viscous jet under gravity force ✓ No electrical force ✓ Jet radius decreases near zero ✓ Velocity and pressure of jet only change during axial direction ✓ Mass and volume control equations and Taylor expansion were applied to predict jet radius.	1996	[224]

TABLE 1.5 *(Continued)*

Researchers	Model	Year	Ref.
Saville, D. A.	Electrohydrodynamic model	1997	[225]
	✓ The hydrodynamic equations of dielectric model were modified.		
	✓ Using dielectric assumption		
	✓ This model can predict drop formation		
	✓ Considering jet as a cylinder (ignoring the diameter reduction)		
	✓ Only for steady-state part of the jet		
Spivak, A.	Spivak and Dzenis model	1998	[150]
Dzenis, Y.	✓ The motions of a viscose fluid jet with lower conductivity were surveyed in an external electric field.		
	✓ Single Newtonian fluid jet		
	✓ The electric field assumed to be uniform and constant, unaffected by the charges carried by the jet		
	✓ Use asymptotic approximation was applied in a long distance from the nozzle.		
	✓ Tangential electric force assumed to be zero.		
	✓ Using nonlinear rheological constitutive equation (Ostwald–de Waele law), nonlinear behavior of fluid jet were investigated.		
Jong Wook	Droplet formation model	2000	[226]
	✓ Droplet formation of charged fluid jet was studied in this model.		
	✓ The ratio of mass, energy and electric charge transition are the most important parameters on droplet formation.		
	✓ Deformation and breakup of droplets were investigated too.		
	✓ Newtonian and non-Newtonian fluids		
	✓ Only for high conductivity and viscose fluids		

TABLE 1.5   *(Continued)*

Researchers	Model	Year	Ref.
Reneker, D. H. Yarin, A. L.	Reneker model ✓ For description of instabilities in viscoelastic jets. ✓ Using molecular chain theory, behavior of polymer chain of spring-bead model in electric field was studied. ✓ Electric force based on electric field cause instability of fluid jet while repulsion force between surface charges make perturbation and bending instability. ✓ The motion paths of these two cases were studied ✓ Governing equations: momentum balance, motion equations for each bead, Maxwell tension and Columbic equations	2000	[227]
Hohman, M. Shin, M.	Stability Theory ✓ This model is based on a dielectric model with some modification for Newtonian fluids. ✓ This model can describe whipping, bending and Rayleigh instabilities and introduced new ballooning instability. ✓ Four motion regions were introduced: dipping mode, spindle mode, oscillating mode, precession mode. ✓ Surface charge density introduced as the most effective parameter on instability formation. ✓ Effect of fluid conductivity and viscosity on nanofibers diameter were discussed. ✓ Steady solutions may be obtained only if the surface charge density at the nozzle is set to zero or a very low value.	2001	[60]

TABLE 1.5 *(Continued)*

Researchers	Model	Year	Ref.
Feng, J. J	✓ Modifying Hohman model	2002	[153]
	✓ For both Newtonian and non-Newtonian fluids		
	✓ Unlike Hohman model, the initial surface charge density was not zero, so the "ballooning instability" did not accrue.		
	✓ Only for steady-state part of the jet		
	✓ Simplifying the electric field equation which Hohman used in order to eliminate ballooning instability.		
Wan–Guo–Pan	Wan–Guo–Pan model	2004	[175]
	✓ They introduced thermo-electro-hydro dynamics model in electrospinning process		
	✓ This model is a modification on Spivak model which mentioned before		
	✓ The governing equations in this model: Modified Maxwell equation, Navier–Stocks equations, and several rheological constitutive equations.		
Ji-Haun	AC-electrospinning model	2005	[172]
	✓ Whipping instability in this model was distinguished as the most effective parameter on uncontrollable deposition of nanofibers.		
	✓ Applying AC current can reduce this instability so make oriented nanofibers.		
	✓ This model found a relationship between axial distance from nozzle and jet diameter.		
	✓ This model also connected ac frequency and jet diameter.		

TABLE 1.5   *(Continued)*

Researchers	Model	Year	Ref.
Roozemond (Eindhoven University and Technology)	Combination of slender body and dielectric model In this model, a new model for viscoelastic jets in electrospinning was presented by combining these two models. All variables were assumed uniform in cross-section of the jet, but they changed in during z direction. Nanofiber diameter can be predicted.	2007	[228]
Wan	Electromagnetic model ✓ Results indicated that the electromagnetic field which made because of electrical field in charged polymeric jet is the most important reason of helix motion of jet during the process.	2012	[229]
Dasri	Dasri model ✓ This model was presented for description of unstable behavior of fluid jet during electrospinning. ✓ This instability causes random deposition of nanofiber on surface of the collector. ✓ This model described dynamic behavior of fluid by combining assumption of Reneker and Spivak models.	2012	[230]

The most frequent numeric mathematical methods which were used in different models are listed in Table 1.6:

TABLE 1.6   Applied numerical methods for electrospinning

Method	Ref.
Relaxation method	[153, 159, 231]
Boundary integral method (boundary element method)	[199, 226]
Semi-inverse method	[159, 172]
(Integral) control-volume formulation	[224]

TABLE 1.6   *(Continued)*

Method	Ref.
Finite element method	[223]
Kutta–Merson method	[232]
Lattice Boltzmann method with finite difference method	[233]

1.11   ELECTROSPINNING SIMULATION

Electrospun polymer nanofibers demonstrate outstanding mechanical and thermodynamic properties as compared with macroscopic-scale structures. These features are attributed to nanofiber microstructure [234–254]. Theoretical modeling predicts the nanostructure formations during electrospinning. This prediction could be verified by various experimental condition and analysis methods that called are simulation. Numerical simulations can be compared with experimental observations as the last evidence[149, 236].

Parametric analysis and accounting complex geometries in simulation of electrospinning are extremely difficult due to the nonlinearity nature in the problem. Therefore, a lot of researches have done to develop an existing electrospinning simulation of viscoelastic liquids [231].

KEYWORDS

- **Electrospinning**
- **Freeze-drying**
- **Laser-based techniques**
- **Molecular dynamics**
- **Momentum balance**
- **Nanocoatings**
- **Nanoparticles**
- **Nanostructures**
- **Proton exchange mat**
- **Quantum dots**

REFERENCES

1. Poole, C. P.; and Owens, F. J.; Introduction to Nanotechnology. New Jersey, Hoboken: Wiely; **2003**, 400 p.
2. Nalwa, H. S.; Nanostructured Materials and Nanotechnology: Concise Edition. Gulf Professional Publishing; **2001**, 324 p.
3. Gleiter, H.; Nanostructured materials: state of the art and perspectives. *Nanostruct. Mater.* **1995**, *6(1)*, 3–14.
4. Wong, Y.; et al. Selected applications of nanotechnology in textiles. *AUTEX Res. J.* **2006**, *6(1)*, 1–8.
5. Yu, B.; and Meyyappan, M.; Nanotechnology: role in emerging nanoelectronics. *Solid-State Electron.* **2006**, *50(4)*, 536–544.
6. Farokhzad, O. C.; and Langer, R.; Impact of nanotechnology on drug delivery. *ACS Nano.* **2009**, *3(1)*, 16–20.
7. Serrano, E.; Rus, G.; and Garcia-Martinez, J.; Nanotechnology for sustainable energy. *Renew. Sust. Energ. Rev.* **2009**, *13(9)*, 2373–2384.
8. Dreher, K. L.; Health and environmental impact of nanotechnology: toxicological assessment of manufactured nanoparticles. *Toxicol. Sci.* **2004**, *77(1)*, 3–5.
9. Bhushan, B.; Introduction to nanotechnology. In: Springer Handbook of Nanotechnology. Springer; **2010**, 1–13.
10. Ratner, D.; and Ratner, M. A.; Nanotechnology and Homeland Security: New Weapons for New Wars. Prentice Hall Professional; **2004**, 145 p.
11. Aricò, A. S.; et al. Nanostructured materials for advanced energy conversion and storage devices. *Nat. Mater.* **2005**, *4(5)*, 366–377.
12. Wang, Z. L.; Nanomaterials for nanoscience and nanotechnology. *Charact. Nanophase Mater.* **2000**, 1–12.
13. Gleiter, H.; Nanostructured materials: basic concepts and microstructure. *Acta Mater.* **2000**, *48(1)*, 1–29.
14. Wang, X.; et al. A general strategy for nanocrystal synthesis. *Nature.* **2005**, *437(7055)*, 121–124.
15. Kelsall, R. W.; et al. Nanoscale Science and Technology. New York: Wiley Online Library; **2005**, 455.
16. Engel, E.; et al. Nanotechnology in regenerative medicine: the materials side. *Trends in Biotechnol.* **2008**, *26(1)*, 39–47.
17. Beachley, V.; and Wen, X.; Polymer nanofibrous structures: fabrication, biofunctionalization, and cell interactions. *Prog. Polym. Sci.* **2010**, *35(7)*, 868–892.
18. Gogotsi, Y.; Nanomaterials Handbook. New york: CRC Press; **2006**, 779.
19. Li, C.; and Chou, T.; A structural mechanics approach for the analysis of carbon nanotubes. *Int. J. Solids Struct.* **2003**, *40(10)*, 2487–2499.
20. Delerue, C.; and Lannoo, M.; Nanostructures: Theory and Modelling. Springer; **2004**, 304 p.
21. Pokropivny, V.; and Skorokhod, V.; Classification of nanostructures by dimensionality and concept of surface forms engineering in nanomaterial science. *Mater. Sci. Eng. C.* **2007**, *27(5)*, 990–993.

22. Balbuena, P.; and Seminario, J. M.; Nanomaterials: Design and Simulation: Design and Simulation. Elsevier; **2006**, *18,* 523.

23. Kawaguchi, T.; and Matsukawa, H.; Numerical study of nanoscale lubrication and friction at solid interfaces. *Mol. Phys.* **2002**, *100(19),* 3161–3166.

24. Ponomarev, S. Y.; Thayer, K. M.; and Beveridge, D. L.; Ion Motions in Molecular Dynamics Simulations on DNA. Proceedings of the National Academy of Sciences of the United States of America. **2004**, *101(41),* 14771–14775.

25. Loss, D.; and DiVincenzo, D. P.; Quantum computation with quantum dots. *Phys. Rev. A.* **1998**, *57(1),* 120–125.

26. Theodosiou, T. C.; and Saravanos, D. A.; Molecular mechanics based finite element for carbon nanotube modeling. *Comput. Model. Eng. Sci.* **2007**, *19(2),* 19–24.

27. Pokropivny, V.; and Skorokhod, V.; New dimensionality classifications of nanostructures. *Phys. E: Low-Dimens. Syst. Nanostruct.* **2008**, *40(7),* 2521–2525.

28. Lieber, C. M.; One-dimensional nanostructures: chemistry, physics & applications. *Solid State Commun.* **1998**, *107(11),* 607–616.

29. Emary, C.; Theory of Nanostructures. New york: Wiley; **2009**, 141.

30. Edelstein, A. S.; and Cammaratra, R. C.; Nanomaterials: Synthesis, Properties and Applications. CRC Press; **1998**.

31. Grzelczak, M.; et al. Directed self-assembly of nanoparticles. *ACS Nano.* **2010**, *4(7),* 3591–3605.

32. Hung, C.; et al. Strain directed assembly of nanoparticle arrays within a semiconductor. *J. Nanopart. Res.* **1999**, *1(3),* 329–347.

33. Wang, L.; and Hong, R.; Synthesis, surface modification and characterisation of nanoparticles. *Polym. Compos.* **2001**, *2,* 13–51.

34. Lai, W.; et al. Synthesis of nanostructured materials by hot and cold plasma. In: *Int. Plasma Chem. Soc.* Orleans, France; **2012**, 5 p.

35. Petermann, N.; et al. Plasma synthesis of nanostructures for improved thermoelectric properties. *J. Phys. D: Appl. Phys.* **2011**, *44(17),* 174034.

36. Ye, Y.; et al. RF Plasma Method. Google Patents: USA; **2001**.

37. Hyeon, T.; Chemical synthesis of magnetic nanoparticles. *Chem. Commun.* **2003**, *8,* 927–934.

38. Galvez, A.; et al. Carbon nanoparticles from laser pyrolysis. *Carbon.* **2002**, *40(15),* 2775–2789.

39. Porterat, D.; Synthesis of Nanoparticles by Laser Pyrolysis. Google Patents: USA; **2012**.

40. Tiwari, J. N.; Tiwari, R. N.; and Kim, K. S.; Zero-dimensional, one-dimensional, two-dimensional and three-dimensional nanostructured materials for advanced electrochemical energy devices. *Prog. Mater. Sci.* **2012**, *57(4),* 724–803.

41. Murray, P. T.; et al. Nanomaterials produced by laser ablation techniques Part I: synthesis and passivation of nanoparticles, in nondestructive evaulation for health monitoring and diagnostics. *Int. Soc. Opt. Phot.* **2006**, 61750–61750.

42. Dolgaev, S. I.; et al. Nanoparticles produced by laser ablation of solids in liquid environment. *Appl. Surf. Sci.* **2002**, *186(1),* 546–551.

43. Becker, M. F.; et al. Metal nanoparticles generated by laser ablation. *Nanostruct. Mater.* **1998**, *10(5),* 853–863.

44. Bonneau, F.; et al. Numerical simulations for description of UV laser interaction with gold nanoparticles embedded in silica. *Appl. Phys. B.* **2004,** *78(3–4),* 447–452.

45. Chen, Y. H.; and Yeh, C. S.; Laser ablation method: use of surfactants to form the dispersed Ag nanoparticles. *Colloids Surf. A: Physicochem. Eng. Asp.* **2002,** *197(1),* 133–139.

46. Andrady, A. L.; Science and Technology of Polymer Nanofibers. Hoboken: John Wiley & Sons Inc.; **2008,** 404 p.

47. Wang, H. S.; Fu, G. D.; and Li, X. S.; Functional polymeric nanofibers from electro-spinning. *Recent Patents Nanotechnol.* **2009,** *3(1),* 21–31.

48. Ramakrishna, S.; An Introduction to Electrospinning and Nanofibers. World Scientific Publishing Company; **2005,** 396 p.

49. Reneker, D. H.; and Chun, I.; Nanometer diameter fibres of polymer, produced by electrospinning. *Nanotechnol.* **1996,** *7(3),* 216.

50. Doshi, J.; and Reneker, D. H.; Electrospinning process and applications of electros-pun fibers. *J. Electrostat.* **1995,** *35(2),* 151–160.

51. Burger, C.; Hsiao, B.; and Chu, B.; Nanofibrous materials and their applications. *Ann. Rev. Mater. Res.* **2006,** *36,* 333–368.

52. Fang, J.; et al. Applications of electrospun nanofibers. *Chin. Sci. Bull.* **2008,** *53(15),* 2265–2286.

53. Ondarcuhu, T.; and Joachim, C.; Drawing a single nanofibre over hundreds of microns. *EPL (Europhys. Lett.)* **1998,** *42(2),* 215.

54. Nain, A. S.; et al. Drawing suspended polymer micro/nanofibers using glass micro-pipettes. *Appl. Phys. Lett.* **2006,** *89(18),* 183105–183105–3.

55. Bajakova, J.; et al. "Drawing"-the production of individual nanofibers by experimental method. In: *Nanoconf.* Brno, Czech Republic, EU; **2011.**

56. Feng, L.; et al. Super hydrophobic surface of aligned polyacrylonitrile nanofibers. *Angew. Chemie.* **2002,** *114(7),* 1269–1271.

57. Delvaux, M.; et al. Chemical and electrochemical synthesis of polyaniline micro-and nano-tubules. *Synthetic Met.* **2000,** *113(3),* 275–280.

58. Barnes, C. P.; et al. Nanofiber technology: designing the next generation of tissue engineering scaffolds. *Adv. Drug Deliv. Rev.* **2007,** *59(14),* 1413–1433.

59. Palmer, L. C.; and Stupp, S. I.; Molecular self-assembly into one-dimensional nano-structures. *Acc. Chem. Res.* **2008,** *41(12),* 1674–1684.

60. Hohman, M. M.; et al. Electrospinning and electrically forced jets. I. Stability theory. *Phys. Fluids.* **2001,** *13,* 2201–2220.

61. Hohman, M. M.; et al. Electrospinning and electrically forced jets. II. Applications. *Phys. Fluids.* **2001,** *13,* 2221.

62. Shin, Y. M.; et al. Experimental characterization of electrospinning: the electrically forced jet and instabilities. *Polym.* **2001,** *42(25),* 9955–9967.

63. Fridrikh, S. V.; et al. Controlling the fiber diameter during electrospinning. *Phys. Rev. Lett.* **2003,** *90(14),* 144502–144502.

64. Yarin, A. L.; Koombhongse, S.; and Reneker, D. H.; Taylor cone and jetting from liquid droplets in electrospinning of nanofibers. *J. Appl. Phys.* **2001,** *90(9),* 4836–4846.

65. Zeleny, J.; The electrical discharge from liquid points, and a hydrostatic method of measuring the electric intensity at their surfaces. *Phys. Rev.* **1914,** *3(2),* 69–91.

66. Reneker, D. H.; et al. Bending instability of electrically charged liquid jets of polymer solutions in electrospinning. *J. Appl. Phys.* **2000,** *87,* 4531–4547.
67. Frenot, A.; and Chronakis, I. S.; Polymer nanofibers assembled by electrospinning. *Current Opinion Colloid Interf. Sci.* **2003,** *8(1),* 64–75.
68. Gilbert, W.; De Magnete Transl. PF Mottelay, Dover, UK. New York: Dover Publications Inc.; **1958,** 366 p.
69. Tucker, N.; et al. The history of the science and technology of electrospinning from 1600 to 1995. J. Eng. Fibers Fabrics. **2012,** *7,* 63–73.
70. Hassounah, I.; Melt Electrospinning of Thermoplastic Polymers. Aachen: Hochschulbibliothek Rheinisch-Westfälische Technischen Hochschule Aachen; **2012,** 650 p.
71. Taylor, G. I.; The scientific papers of sir geoffrey ingram taylor. *Mech. Fluids.* **1971,** *4.*
72. Yeo, L. Y.; and Friend, J. R.; Electrospinning carbon nanotube polymer composite nanofibers. *J. Exp. Nanosci.* **2006,** *1(2),* 177–209.
73. Bhardwaj, N.; and Kundu, S. C.; Electrospinning: a fascinating fiber fabrication technique. *Biotechnol. Adv.* **2010,** *28(3),* 325–347.
74. Huang, Z. M.; et al. A review on polymer nanofibers by electrospinning and their applications in nanocomposites. *Compos. Sci. Technol.* **2003,** *63(15),* 2223–2253.
75. Haghi, A. K.; Electrospinning of Nanofibers in Textiles. North Calorina: Apple Academic Press Inc; **2011,** 132.
76. Bhattacharjee, P.; Clayton, V.; and Rutledge, A. G.; Electrospinning and Polymer Nanofibers: Process Fundamentals. In: Comprehensive Biomaterials. Elsevier. **2011,** 497–512 p.
77. Garg, K.; and Bowlin, G. L.; Electrospinning jets and nanofibrous structures. *Biomicrofluidics.* **2011,** *5,* 13403–13421.
78. Angammana, C. J.; and Jayaram, S. H.; A Theoretical Understanding of the Physical Mechanisms of Electrospinning. In: Proc. ESA Annual Meeting on Electrostatics. Cleveland OH: Case Western Reserve University; **2011,** 1–9 p.
79. Reneker, D. H.; and Yarin, A. L.; Electrospinning jets and polymer nanofibers. *Polym.* **2008,** *49(10),* 2387–2425.
80. Deitzel, J.; et al. The effect of processing variables on the morphology of electrospun nanofibers and textiles. *Polym.* **2001,** *42(1),* 261–272.
81. Rutledge, G. C.; and Fridrikh, S. V.; Formation of fibers by electrospinning. *Adv. Drug Deliv. Rev.* **2007,** *59(14),* 1384–1391.
82. De Vrieze, S.; et al. The effect of temperature and humidity on electrospinning. *J. Mater. Sci.* **2009,** *44(5),* 1357–1362.
83. Kumar, P.; Effect of colletor on electrospinning to fabricate aligned nanofiber. In: Department of Biotechnology and Medical Engineering. Rourkela: National Institute of Technology Rourkela; **2012,** 88 p.
84. Sanchez, C.; Arribart, H.; and Guille, M.; Biomimetism and bioinspiration as tools for the design of innovative materials and systems. *Nat. Mater.* **2005,** *4(4),* 277–288.
85. Ko, F.; et al. Electrospinning of continuous carbon nanotube-filled nanofiber yarns. *Adv. Mater.* **2003,** *15(14),* 1161–1165.
86. Stuart, M.; et al. Emerging applications of stimuli-responsive polymer materials. *Nat. Mater.* **2010,** *9(2),* 101–113.

87. Gao, W.; Chan, J.; and Farokhzad, O.; pH-responsive nanoparticles for drug delivery. Mol. Pharmaceutics. **2010**, *7(6),* 1913–1920.

88. Li, Y.; et al. Stimulus-responsive polymeric nanoparticles for biomedical applications. *Sci. China Chem.* **2010**, *53(3),* 447–457.

89. Tirelli, N.; (Bio) Responsive nanoparticles. *Current Opinion Colloid Interf. Sci.* **2006**, *11(4),* 210–216.

90. Bonini, M.; et al. A new way to prepare nanostructured materials: flame spraying of microemulsions. *J. Phys. Chem. B.* **2002**, *106(24),* 6178–6183.

91. Thierry, B.; et al. Nanocoatings onto arteries via layer-by-layer deposition: toward the in vivo repair of damaged blood vessels. *J. Am. Chem. Soc.* **2003**, *125(25),* 7494–7495.

92. Andrady, A.; Science and Technology of Polymer Nanofibers. Wiley. com. **2008**.

93. Carroll, C. P.; et al. Nanofibers from electrically driven viscoelastic jets: modeling and experiments. *Korea-Aust. Rheol. J.* **2008**, *20(3),* 153–164.

94. Zhao, Y.; and Jiang, L.; Hollow micro/nanomaterials with multilevel interior structures. *Adv. Mater.* **2009**, *21(36),* 3621–3638.

95. Carroll, C. P.; The development of a comprehensive simulation model for electrospinning. Cornell University; **2009**, *70,* 300 p.

96. Song, Y. S.; and Youn, J. R.; Modeling of rheological behavior of nanocomposites by Brownian dynamics simulation. *Korea–Aust. Rheol. J.* **2004**, *16(4),* 201–212.

97. Dror, Y.; et al. Carbon nanotubes embedded in oriented polymer nanofibers by electrospinning. *Langmuir.* **2003**, *19(17),* 7012–7020.

98. Gates, T.; et al. Computational materials: multi-scale modeling and simulation of nanostructured materials. *Compos. Sci. Technol.* **2005**, *65(15),* 2416–2434.

99. Agic, A.; Multiscale mechanical phenomena in electrospun carbon nanotube composites. *J. Appl. Polym. Sci.* **2008**, *108(2),* 1191–1200.

100. Teo, W.; and Ramakrishna, S.; Electrospun nanofibers as a platform for multifunctional, hierarchically organized nanocomposite. *Compos. Sci. Technol.* **2009**, *69(11),* 1804–1817.

101. Silling, S.; and Bobaru, F.; Peridynamic modeling of membranes and fibers. *Int. J. Non-Linear Mech.* **2005**, *40(2),* 395–409.

102. Berhan, L.; et al. Mechanical properties of nanotube sheets: alterations in joint morphology and achievable moduli in manufacturable materials. *J. Appl. Phys.* **2004**, *95(8),* 4335–4345.

103. Heyden, S.; Network Modelling for the Evaluation of Mechanical Properties of Cellulose Fibre Fluff. Lund University; **2000**.

104. Collins, A. J.; et al. The value of modeling and simulation standards. Virginia Modeling, Analysis and Simulation Center. Virginia: Old Dominion University; **2011**, 1–8 p.

105. Kuwabara, S.; The forces experienced by randomly distributed parallel circular cylinders or spheres in a viscous flow at small Reynolds numbers. *J. Phys. Soc. Jpn.* **1959**, *14,* 527.

106. Brown, R.; Air Filtration: An Integrated Approach to the Theory and Applications of Fibrous Filters. New York: Pergamon Press New York; **1993**.

107. Buysse, W. M.; et al. A 2D model for the electrospinning process. In: Department of Mechanical Engineering. Eindhoven: Eindhoven University of Technology; **2008**, 75.

108. Ante, A.; and Budimir, M.; Design Multifunctional Product by Nanostructures. Sciyo. com. **2010,** 27.

109. Jackson, G.; and James, D.; The permeability of fibrous porous media. *Can. J. Chem. Eng.* **1986,** *64(3),* 364–374.

110. Sundmacher, K.; Fuel cell engineering: toward the design of efficient electrochemical power plants. *Ind. Eng. Chem. Res.* **2010,** *49(21),* 10159–10182.

111. Kim, Y.; et al. Electrospun bimetallic nanowires of PtRh and PtRu with compositional variation for methanol electrooxidation. *Electrochem. Commun.* **2008,** *10(7),* 1016–1019.

112. Kim, H.; et al. Pt and PtRh nanowire electrocatalysts for cyclohexane-fueled polymer electrolyte membrane fuel cell. *Electrochem. Commun.* **2009,** *11(2),* 446–449.

113. Formo, E.; et al. Functionalization of electrospun TiO2 nanofibers with Pt nanoparticles and nanowires for catalytic applications. *Nano Lett.* **2008,** *8(2),* 668–672.

114. Xuyen, N.; et al. Hydrolysis-induced immobilization of Pt (acac) 2 on polyimidebased carbon nanofiber mat and formation of Pt nanoparticles. *J. Mater. Chem.* **2009,** *19(9),* 1283–1288.

115. Lee, K.; et al. Nafion nanofiber membranes. *ECS Transact.* **2009,** *25(1),* 1451–1458.

116. Qu, H.; Wei, S.; and Guo, Z.; Coaxial electrospun nanostructures and their applications. *J. Mater. Chem. A.* **2013,** *1(38),* 11513–11528.

117. Thavasi, V.; Singh, G.; and Ramakrishna, S.; Electrospun nanofibers in energy and environmental applications. *Energ. Environ. Sci.* **2008,** *1(2),* 205–221.

118. Dersch, R.; et al. Nanoprocessing of polymers: applications in medicine, sensors, catalysis, photonics. *Polym. Adv. Technol.* **2005,** *16(2–3),* 276–282.

119. Yih, T.; and Al-Fandi, M.; Engineered nanoparticles as precise drug delivery systems. *J. Cell. Biochem.* **2006,** *97(6),* 1184–1190.

120. Kenawy, E.; et al. Release of tetracycline hydrochloride from electrospun poly (ethylene-co-vinylacetate), poly (lactic acid), and a blend. *J. Controlled Release.* **2002,** *81(1),* 57–64.

121. Verreck, G.; et al. Incorporation of drugs in an amorphous state into electrospun nanofibers composed of a water-insoluble, nonbiodegradable polymer. *J. Controlled Release.* **2003,** *92(3),* 349–360.

122. Zeng, J.; et al. Biodegradable electrospun fibers for drug delivery. *J. Controlled Release.* **2003,** *92(3),* 227–231.

123. Luu, Y.; et al. Development of a nanostructured DNA delivery scaffold via electrospinning of PLGA and PLA–PEG block copolymers. *J. Controlled Release.* **2003,** *89(2),* 341–353.

124. Zong, X.; et al. Structure and process relationship of electrospun bioabsorbable nanofiber membranes. *Polym.* **2002,** *43(16),* 4403–4412.

125. Yu, D.; et al. PVP nanofibers prepared using co-axial electrospinning with salt solution as sheath fluid. *Mater. Lett.* **2012,** *67(1),* 78–80.

126. Verreck, G.; et al. Preparation and characterization of nanofibers containing amorphous drug dispersions generated by electrostatic spinning. *Pharm. Res.* **2003,** *20(5),* 810–817.

127. Jiang, H.; et al. Preparation and characterization of ibuprofen-loaded poly (lactide-co-glycolide)/poly (ethylene glycol)-g-chitosan electrospun membranes. *J. Biomater. Sci. Polym. Edn.* **2004,** *15(3),* 279–296.
128. Yang, D.; Li, Y.; and Nie, J.; Preparation of gelatin/PVA nanofibers and their potential application in controlled release of drugs. *Carbohydr. Polym.* **2007,** *69(3),* 538–543.
129. Kim, K.; et al. Incorporation and controlled release of a hydrophilic antibiotic using poly (lactide-co-glycolide)-based electrospun nanofibrous scaffolds. *J. Controlled Release.* **2004,** *98(1),* 47–56.
130. Xu, X.; et al. Ultrafine medicated fibers electrospun from W/O emulsions. *J. Controlled Release.* **2005,** *108(1),* 33–42.
131. Zeng, J.; et al. Poly (vinyl alcohol) nanofibers by electrospinning as a protein delivery system and the retardation of enzyme release by additional polymer coatings. *Biomacromol.* **2005,** *6(3),* 1484–1488.
132. Yun, J.; et al. Effect of oxyfluorination on electromagnetic interference shielding behavior of MWCNT/PVA/PAAc composite microcapsules. *Euro. Polym. J.* **2010,** *46(5),* 900–909.
133. Jiang, H.; et al. A facile technique to prepare biodegradable coaxial electrospun nanofibers for controlled release of bioactive agents. *J. Controlled Release.* **2005,** *108(2),* 237–243.
134. He, C.; Huang, Z.; and Han, X.; Fabrication of drug-loaded electrospun aligned fibrous threads for suture applications. *J. Biomed. Mater. Res. Part A.* **2009,** *89(1),* 80–95.
135. Qi, R.; et al. Electrospun poly (lactic-co-glycolic acid)/halloysite nanotube composite nanofibers for drug encapsulation and sustained release. *J. Mater. Chem.* **2010,** *20(47),* 10622–10629.
136. Reneker, D. H.; et al. Electrospinning of nanofibers from polymer solutions and melts. *Adv. Appl. Mech.* **2007,** *41,* 343–346.
137. Haghi, A. K.; and Zaikov, G.; Advances in Nanofibre Research. Smithers Rapra Technology; **2012,** 194 p.
138. Maghsoodloo, S.; et al. A detailed review on mathematical modeling of electrospun nanofibers. *Polym. Res. J.* **2012,** *6,* 361–379.
139. Fritzson, P.; Principles of object-oriented modeling and simulation with Modelica 2.1. Wiley-IEEE Press; **2010.**
140. Robinson, S.; Simulation: The Practice of Model Development and Use. Wiley; **2004,** 722 p.
141. Carson, I. I.; and John, S.; Introduction to modeling and simulation. In: Proceedings of the 36th Conference on Winter Simulation. Washington, DC: Winter Simulation Conference; **2004,** 9–16 p.
142. Banks, J.; Handbook of Simulation. Wiley Online Library; **1998,** 342 p.
143. Pritsker, A. B.; and Alan, B.; Principles of Simulation Modeling. New York: Wiley; **1998,** 426 p.
144. Yu, J. H.; Fridrikh, S. V.; and Rutledge, G. C.; The role of elasticity in the formation of electrospun fibers. *Polym.* **2006,** *47(13),* 4789–4797.
145. Han, T.; Yarin, A. L.; and Reneker, D. H.; Viscoelastic electrospun jets: initial stresses and elongational rheometry. *Polym.* **2008,** *49(6),* 1651–1658.

146. Bhattacharjee, P. K.; et al. Extensional stress growth and stress relaxation in entangled polymer solutions. *J. Rheol.* **2003,** *47,* 269–290.

147. Paruchuri, S.; and Brenner, M. P.; Splitting of a liquid jet. *Phys. Rev. Lett.* **2007,** *98(13),* 134502–134504.

148. Ganan-Calvo, A. M.; On the theory of electrohydrodynamically driven capillary jets. *J. Fluid Mech.* **1997,** *335,* 165–188.

149. Liu, L.; and Dzenis, Y. A.; Simulation of electrospun nanofibre deposition on stationary and moving substrates. *Micro Nano Lett.* **2011,** *6(6),* 408–411.

150. Spivak, A. F.; and Dzenis, Y. A.; Asymptotic decay of radius of a weakly conductive viscous jet in an external electric field. *Appl. Phys. Lett.* **1998,** *73(21),* 3067–3069.

151. Jaworek, A.; and Krupa, A.; Classification of the modes of EHD spraying. *J. Aerosol Sci.* **1999,** *30(7),* 873–893.

152. Senador, A. E., Shaw, M. T.; and Mather, P. T.; Electrospinning of polymeric nanofibers: analysis of jet formation. In: *Mater. Res. Soc.* California, USA: Cambridge Univ Press; **2000,** 11 p.

153. Feng, J. J.; The stretching of an electrified non-Newtonian jet: a model for electrospinning. *Phys. Fluids.* **2002,** *14(11),* 3912–3927.

154. Feng, J. J.; Stretching of a straight electrically charged viscoelastic jet. *J. Non-Newtonian Fluid Mech.* **2003,** *116(1),* 55–70.

155. Spivak, A. F.; Dzenis, Y. A.; and Reneker, D. H.; A model of steady state jet in the electrospinning process. *Mech. Res. Commun.* **2000,** *27(1),* 37–42.

156. Yarin, A. L.; Koombhongse, S.; and Reneker, D. H.; Bending instability in electrospinning of nanofibers. *J. Appl. Phys.* **2001,** *89,* 3018.

157. Gradoń, L.; Principles of momentum, mass and energy balances. *Chem. Eng. Chem. Process Technol. 1,* 1–6.

158. Bird, R. B.; Stewart, W. E.; and Lightfoot, E. N.; Transport Phenomena. New York: Wiley & Sons, Incorporated, John; **1960,** *2,* 808.

159. Peters, G. W. M.; Hulsen, M. A.; and Solberg, R. H. M.; A Model for Electrospinning Viscoelastic Fluids, in Department of Mechanical Engineering. Eindhoven: Eindhoven University of Technology; **2007,** 26 p.

160. Whitaker, R. D.; An historical note on the conservation of mass. *J. Chem. Educ.* **1975,** *52(10),* 658.

161. He, J. H.; et al. Mathematical models for continuous electrospun nanofibers and electrospun nanoporous microspheres. *Polym. Int.* **2007,** *56(11),* 1323–1329.

162. Xu, L.; Liu, F.; and Faraz, N.; Theoretical model for the electrospinning nanoporous materials process. *Comput. Math. Appl.* **2012,** *64(5),* 1017–1021.

163. Heilbron, J. L.; Electricity in the 17th and 18th Century: A Study of Early Modern Physics. University of California Press; **1979,** 437 p.

164. Orito, S.; and Yoshimura, M.; Can the universe be charged? *Phys. Rev. Lett.* **1985,** *54(22),* 2457–2460.

165. Karra, S.; Modeling electrospinning process and a numerical scheme using Lattice Boltzmann method to simulate viscoelastic fluid flows. In: Indian Institute of Technology. Chennai: Texas A & M University; **2007,** 60 p.

166. Hou, S. H.; and Chan, C. K.; Momentum equation for straight electrically charged jet. *Appl. Math. Mech.* **2011,** *32(12),* 1515–1524.

167. Maxwell, J. C.; Electrical Research of the Honorable Henry Cavendish, 426, in Cambridge University Press, Cambridge, Editor. Cambridge, UK: Cambridge University Press; **1878**.
168. Vught, R. V.; Simulating the dynamical behaviour of electrospinning processes. In: Department of Mechanical Engineering. Eindhoven: Eindhoven University of Technology; **2010, 68**.
169. Jeans, J. H.; The Mathematical Theory of Electricity and Magnetism. London: Cambridge University Press; **1927, 536**.
170. Truesdell, C.; and Noll, W.; The Non-Linear Field Theories of Mechanics. Springer; **2004**, 579 p.
171. Roylance, D.; Constitutive equations. In: Lecture Notes. Department of Materials Science and Engineering. Cambridge: Massachusetts Institute of Technology; **2000**, 10 p.
172. He, J. H.; Wu, Y.; and Pang, N.; A mathematical model for preparation by AC-electrospinning process. *Int. J. Nonlinear Sci. Numer. Simul.* **2005**, *6(3)*, 243–248.
173. Little, R. W.; Elasticity. Courier Dover Publications; **1999**, 431.
174. Clauset, A.; Shalizi, C. R.; and Newman, M. E. J.; Power-law distributions in empirical data. *SIAM Rev.* **2009**, *51(4), 661*–703.
175. Wan, Y.; Guo, Q.; and Pan, N.; Thermo-electro-hydrodynamic model for electrospinning process. *Int. J. Nonlinear Sci. Numer. Simul.* **2004**, *5(1)*, 5–8.
176. Giesekus, H.; Die elastizität von flüssigkeiten. *Rheol. Acta.* **1966**, *5(1)*, 29–35.
177. Giesekus, H.; The physical meaning of Weissenberg's hypothesis with regard to the second normal-stress difference. In: The Karl Weissenberg 80th Birthday Celebration Essays. Eds. Harris, J.; and Weissenberg, K.; East African Literature Bureau; **1973**, 103–112 p.
178. Wiest, J. M.; A differential constitutive equation for polymer melts. *Rheol. Acta.* **1989**, *28(1)*, 4–12.
179. Bird, R. B.; and Wiest, J. M.; Constitutive equations for polymeric liquids. *Ann. Rev. Fluid Mech.* **1995**, *27(1)*, 169–193.
180. Giesekus, H.; A simple constitutive equation for polymer fluids based on the concept of deformation-dependent tensorial mobility. *J. Non-Newtonian Fluid Mech.* **1982**, *11(1)*, 69–109.
181. Oliveira, P. J.; On the numerical implementation of nonlinear viscoelastic models in a finite-volume method. *Numer. Heat Transfer: Part B: Fundam.* **2001**, *40(4)*, 283–301.
182. Simhambhatla, M.; and Leonov, A. I.; On the rheological modeling of viscoelastic polymer liquids with stable constitutive equations. *Rheol. Acta.* **1995**, *34(3)*, 259–273.
183. Giesekus, H.; A unified approach to a variety of constitutive models for polymer fluids based on the concept of configuration-dependent molecular mobility. *Rheol. Acta.* **1982**, *21(4–5)*, 366–375.
184. Eringen, A. C.; and Maugin, G. A.; Electrohydrodynamics. In: Electrodynamics of Continua II. Springer; **1990**, 551–573 p.
185. Hutter, K.; Electrodynamics of continua (A. Cemal Eringen and Gerard A. Maugin). *SIAM Rev.* **1991**, *33(2)*, 315–320.

186. Kröger, M.; Simple models for complex nonequilibrium fluids. *Phys. Rep.* **2004,** *390(6),* 453–551.

187. Denn, M. M.; Issues in viscoelastic fluid mechanics. *Ann. Rev. Fluid Mech.* **1990,** *22(1),* 13–32.

188. Rossky, P. J.; Doll, J. D.; and Friedman, H. L.; Brownian dynamics as smart monte carlo simulation. *J. Chem. Phys.* **1978,** *69,* 4628–4633.

189. Chen, J. C.; and Kim, A. S.; Brownian dynamics, molecular dynamics, and monte carlo modeling of colloidal systems. *Adv. Colloid Interf. Sci.* **2004,** *112(1),* 159–173.

190. Pasini, P.; and Zannoni, C.; Computer Simulations of Liquid Crystals and Polymers. Erice: Springerl; **2005,** *177,* 380 p.

191. Zhang, H.; and Zhang, P.; Local existence for the FENE-dumbbell model of polymeric fluids. *Arch. Rat. Mech. Anal.* **2006,** *181(2),* 373–400.

192. Isihara, A.; Theory of high polymer solutions (the dumbbell model). *J. Chem. Phys.* **1951,** *19,* 397–343.

193. Masmoudi, N.; Well-posedness for the FENE dumbbell model of polymeric flows. *Commun. Pure Appl. Math.* **2008,** *61(12),* 1685–1714.

194. Stockmayer, W. H.; et al. Dynamic properties of solutions. Models for chain molecule dynamics in dilute solution. *Discuss. Faraday Soc.* **1970,** *49,* 182–192.

195. Graham, R. S.; et al. Microscopic theory of linear, entangled polymer chains under rapid deformation including chain stretch and convective constraint release. *J. Rheol.* **2003,** *47,* 1171–1200.

196. Gupta, R. K.; Kennel, E.; and Kim, K. S.; Polymer Nanocomposites Handbook. CRC Press; **2010.**

197. Marrucci, G.; The free energy constitutive equation for polymer solutions from the dumbbell model. *J. Rheol.* **1972,** *16,* 321–331.

198. Reneker, D. H.; et al. Electrospinning of nanofibers from polymer solutions and melts. *Adv. Appl. Mech.* **2007,** *41,* 43–195.

199. Kowalewski, T. A.; Barral, S.; and Kowalczyk, T.; Modeling electrospinning of nanofibers. In: IUTAM Symposium on Modelling Nanomaterials and Nanosystems. Aalborg, Denmark: Springer:; **2009,** 279–292 p.

200. Macosko, C. W.; Rheology: principles, measurements, and applications. Poughkeepsie. Newyork: Wiley-VCH; **1994,** 578 p.

201. Kowalewski, T. A.; Blonski, S.; and Barral, S.; Experiments and modelling of electrospinning process. *Tech. Sci.* **2005,** *53(4),* 385–394.

202. Ma, W. K. A.; et al. Rheological modeling of carbon nanotube aggregate suspensions. *J. Rheol.* **2008,** *52,* 1311–1330.

203. Buysse, W. M.; A 2D Model for the Electrospinning Process, in Department of Mechanical Engineering. Eindhoven: Eindhoven University of Technology; **2008,** 71 p.

204. Silling, S. A.; and Bobaru, F.; Peridynamic modeling of membranes and fibers. *Int. J. Non-Linear Mech.* **2005,** *40(2),* 395–409.

205. Teo, W. E.; and Ramakrishna, S.; Electrospun nanofibers as a platform for multifunctional, hierarchically organized nanocomposite. *Compos. Sci. Technol.* **2009,** *69(11),* 1804–1817.

206. Wu, X.; and Dzenis, Y. A.; Elasticity of planar fiber networks. *J. Appl. Phys.* **2005,** *98(9),* 93501.

207. Tatlier, M.; and Berhan, L.; Modelling the negative poisson's ratio of compressed fused fibre networks. *Phys. Status Solidi (b).* **2009,** *246(9),* 2018–2024.
208. Kuipers, B.; Qualitative Reasoning: Modeling and Simulation with Incomplete Knowledge. The MIT Press; **1994,** 554 p.
209. West, B. J.; Comments on the renormalization group, scaling and measures of complexity. *Chaos Solitons Fractals.* **2004,** *20(1),* 33–44.
210. De Gennes, P. G.; and Witten, T. A.; Scaling Concepts in Polymer Physics. Cornell University Press; **1980,** 324 p.
211. He, J. H.; and Liu, H. M.; Variational approach to nonlinear problems and a review on mathematical model of electrospinning. *Nonlinear Anal.* **2005,** *63,* e919–e929.
212. He, J. H.; Wan, Y. Q.; and Yu, J. Y.; Allometric scaling and instability in electrospinning. *Int. J. Nonlinear Sci. Numer. Simul.* **2004,** *5(3),* 243–252.
213. He, J. H.; Wan, Y. Q.; and Yu, J. Y.; Allometric scaling and instability in electrospinning. *Int. J. Nonlinear Sci. Numer. Simul.* **2004,** *5,* 243–252.
214. He, J. H.; and Wan, Y. Q.; Allometric scaling for voltage and current in electrospinning. *Polym.* **2004,** *45,* 6731–6734.
215. He, J. H.; Wan, Y. Q.; and Yu, J. Y.; Scaling law in electrospinning: relationship between electric current and solution flow rate. *Polym.* **2005,** *46,* 2799–2801.
216. He, J. H.; Wanc, Y. Q.; and Yuc, J. Y.; Application of vibration technology to polymer electrospinning. *Int. J. Nonlinear Sci. Numer. Simul.* **2004,** *5(3),* 253–262.
217. Kessick, R.; Fenn, J.; and Tepper, G.; The use of AC potentials in electrospraying and electrospinning processes. *Polym.* **2004,** *45(9),* 2981–2984.
218. Boucher, D. F.; and Alves, G. E.; Dimensionless Numbers. Part 1 and 2. **1959**.
219. Ipsen, D. C.; Units Dimensions and Dimensionless Numbers. New York: McGraw Hill Book Company Inc; **1960,** 466 p.
220. Langhaar, H. L.; Dimensional Analysis and Theory of Models. New York: Wiley; **1951,** *2,* 166
221. McKinley, G. H.; Dimensionless groups for understanding free surface flows of complex fluids. *Bull. Soc. Rheol.* **2005,** *2005,* 6–9.
222. Carroll, C. P.; et al. Nanofibers from electrically driven viscoelastic jets: modeling and experiments. *Korea-Aust. Rheol. J.* **2008,** *20(3),* 153–164.
223. Saville, D.; Electrohydrodynamics: the Taylor-Melcher leaky dielectric model. *Ann. Rev. Fluid Mech.* **1997,** *29(1),* 27–64.
224. Ramos, J. I.; Force fields on inviscid, slender, annular liquid. *Int. J. Numer. Methods Fluids.* **1996,** *23,* 221–239.
225. Saville, D. A.; Electrohydrodynamics: the Taylor-Melcher leaky dielectric model. *Ann. Rev. Fluid Mech.* **1997,** *29(1),* 27–64.
226. Senador, A. E.; Shaw, M. T.; and Mather, P. T.; Electrospinning of polymeric nanofibers: analysis of jet formation. In: MRS Proceedings. Cambridge Univ Press; **2000**.
227. Reneker, D. H.; et al. Bending instability of electrically charged liquid jets of polymer solutions in electrospinning. *J. Appl. Phys.* **2000,** *87,* 4531.
228. Peters, G.; Hulsen, M.; and Solberg, R.; A Model for Electrospinning Viscoelastic Fluids.
229. Wan, Y.; et al. Modeling and simulation of the electrospinning jet with archimedean spiral. *Adv. Sci. Lett.* **2012,** *10(1),* 590–592.

230. Dasri, T.; Mathematical models of bead-spring jets during electrospinning for fabrication of nanofibers. *Walailak J. Sci. Technol.* **2012,** 9.

231. Solberg, R. H. M.; Position-Controlled Deposition for Electrospinning. Eindhoven: Eindhoven University of Technology; **2007,** 75 p.

232. Holzmeister, A.; Yarin, A. L.; and Wendorff, J. H.; Barb formation in electrospinning: experimental and theoretical investigations. *Polym.* **2010,** *51(12),* 2769–2778.

233. Karra, S.; Modeling electrospinning process and a numerical scheme using Lattice Boltzmann method to simulate viscoelastic fluid flows. **2012.**

234. Arinstein, A.; et al. Effect of supramolecular structure on polymer nanofibre elasticity. *Nat. Nanotechnol.* **2007,** *2(1),* 59–62.

235. Lu, C.; et al. Computer simulation of electrospinning. Part I. Effect of solvent in electrospinning. *Polym.* **2006,** *47(3),* 915–921.

236. Greenfeld, I.; et al. Polymer dynamics in semidilute solution during electrospinning: a simple model and experimental observations. *Phys. Rev.* **2011,** *84(4),* 41806–41815.

237. Ly, H. V.; and Tran, H. T.; Modeling and control of physical processes using proper orthogonal decomposition. *Math. Comput. Modell.* **2001,** *33(1),* 223–236.

238. Peiró, J.; and Sherwin, S.; Finite difference, finite element and finite volume methods for partial differential equations. In: Handbook of Materials Modeling. London: Springer; **2005,** 2415–2446 p.

239. Kitano, H.; Computational systems biology. *Nature.* **2002,** *420(6912),* 206–210.

240. Gerald, C. F.; and Wheatley, P. O.; Applied Numerical Analysis. Ed. 7th. Addison-Wesley; **2007,** 624 p.

241. Burden, R. L.; and Faires, J. D.; Numerical Analysis. Thomson Brooks/Cole; **2005,** *8,* 850 p.

242. Lawrence, C. E.; Partial Differential Equations. American Mathematical Society; **2010,** 749 p.

243. Quarteroni, A.; Quarteroni, A. M.; and Valli, A.; Numerical Approximation of Partial Differential Equations. Springer; **2008,** *23,* 544 p.

244. Butcher, J. C.; A history of Runge-Kutta methods. *Appl. Numer. Math.* **1996,** *20(3),* 247–260.

245. Cartwright, J. H. E.; and Piro, O.; The dynamics of Runge–Kutta methods. *Int. J. Bifurcat. Chaos.* **1992,** *2(03),* 427–449.

246. Zingg, D. W.; and Chisholm, T. T.; Runge–Kutta methods for linear ordinary differential equations. *Appl. Numer. Math.* **1999,** *31(2),* 227–238.

247. Butcher, J. C.; The Numerical Analysis of Ordinary Differential Equations: Runge-Kutta and General Linear Methods. Wiley-Interscience; **1987,** 512 p.

248. Reznik, S. N.; et al. Evolution of a compound droplet attached to a core-shell nozzle under the action of a strong electric field. *Phys. Fluids.* **2006,** *18(6),* 062101–062101–13.

249. Reznik, S. N.; et al. Transient and steady shapes of droplets attached to a surface in a strong electric field. *J. Fluid Mech.* **2004,** *516,* 349–377.

250. Donea, J.; and Huerta, A.; Finite Element Methods for Flow Problems. Wiley. com. **2003,** 362 p.

251. Zienkiewicz, O. C.; and Taylor, R. L.; The Finite Element Method: Solid Mechanics. Butterworth-Heinemann; **2000,** *2,* 459 p.

252. Brenner, S. C.; and Scott, L. R.; The Mathematical Theory of Finite Element Methods. Springer; **2008,** *15,* 397 p.
253. Bathe, K. J.; Finite Element Procedures. Prentice Hall Englewood Cliffs; **1996,** *2,* 1037 p.
254. Reddy, J. N.; An Introduction to the Finite Element Method. New York: McGraw-Hill; **2006,** *2,* 912 p.

CHAPTER 2

ALUMINIUM-COATED POLYMER FILMS AS INFRARED LIGHT SHIELDS FOR FOOD PACKING

ESEN ARKIŞ and DEVRİM BALKÖSE

CONTENTS

2.1 INTRODUCTION

The protection from harmful effects caused by infrared radiation on foods can be made by using infrared shielding packing materials. Nano-oxides with the surface effect, small-size effect, quantum-size effect, macroscopic quantum tunneling effect, and other special properties can be used in preparation of infrared shielding coating, absorbing coatings, conductive coatings, insulation coatings, and so on [1].

Polymer coatings on the surfaces are very easily degraded because of the infrared light. The life of the polymer coating could be extended if they are coated with an infrared light reflecting or absorbing layer. Aluminum coatings were used for this purpose in many applications [2]. Aluminum oxide coating on polymers is also used for protection of the polymer layer. Covalent bonds are responsible for adhesion of aluminum oxide to polymer surface. Nano Al_2O_3 particles have a wide absorption band in the infrared band, and they have been used widely in the paint, military, scientific and industrial application [2]. Resonance absorption of the infrared radiation may take place at extremely thin metal coatings by reducing the overall temperature of the heat shield [3].

Polymer surfaces are modified by plasma techniques for interfacial enhancement [4–10]. Au, Ag, Pd, Cu, and Ni were coated on poly (methylmethacrylate) (PMMA) by barrel technique [11]. Coating of Al alloys on PET was compared with Ti layer under Al alloys [12]. Thin aluminum oxide coatings have been deposited on various uncoated papers, polymer-coated papers, and plain polymer films using atomic layer deposition technique [13]. The isotactic polypropylene (iPP) and Al composite is widely used as television cable electromagnetic shielding materials [14]. The reflection of infrared light depends on geometry of the surface of aluminum. The reflection of infrared light was reduced when the height of the triangular aluminum gratings were reduced [15]. Al coating obtained by roll-to-roll coating on polypropylene was polycrystalline with a grain size of 20–70 nm [16].

The morphology, order, light transmittance, and water vapor permeability of the Al-coated polypropylene films were reported in a previous publication [17]. The films did not transmit light in the UV and visible region of the light spectrum, and Al coating reduced the water vapor permeation through the films [17].

The objective of this study is to study the absorption and reflection of infrared light by the uncoated and Al-coated surfaces of polymer films using transmission and reflection techniques. For this purpose, two commercial biaxially oriented polypropylene film, a cast polypropylene film, a milk cover and a chocolate coating material with their one surface coated with a thin aluminum layer were examined by infrared spectroscopy. One surface of the each film was coated with a thin aluminum layer.

2.2 EXPERIMENTAL

2.2.1 MATERIALS

Infrared light absorption and reflection of the uncoated and Al-coated surfaces of polymer films were investigated in the present study. Two commercial biaxially oriented polypropylene samples coated with Al by physical vapor deposition technique were examined. The samples called commercial film 1 and commercial film 2 were provided by POLINAS and POLIBAK companies, respectively. They were 16 µm and 19 µm thick, and their one surface was coated with Al after a corona discharge treatment.

An aluminum-coated polypropylene film with 50 µm thickness that was previously prepared by [17] was also examined. Al film on polypropylene was deposited by the high vacuum magnetron sputtering system having four guns. Argon gas (99.9 wt percent) was used to create plasma. Polypropylene film fixed on glass substrates was coated by condensation of Al atoms sputtered from an Al target. In the sputtering process, 20 W DC power and 20 mA current were applied to pure Al target. The Al coating obtained by magnetron sputtering method was 98–131 nm thickness and formed by small Al particles having 22–29 nm grain sizes [18]. Commercial milk and chocolate packaging materials were also investigated.

2.2.2 METHODS

The surface morphology of the films was examined by scanning electron microscopy. FEI QUANTA FEG-250 SEM was used for this purpose. The thickness of Al coating on the surface of the films was measured from the fractured brittle Al layer obtained by stretching the films.

The transmission infrared spectra of the films at 20°C were taken with Excalibur DIGILAB FTS 3000 MX-type Fourier transform infrared spectrophotometer with a resolution of 4 cm^{-1}. DTGS-type detector was used for all measurements. The transmission spectra of the films were obtained by placing the films in two different positions. The incident infrared light first strikes to either Al coated or uncoated surface in these positions. In transmission spectrum, the light that passes through the sample is measured. The grazing angle specular reflectance accessory with 80° (Pike Technologies), in the reflection–absorption mode was used to obtain specular reflectance spectra. Gold-coated glass was used as a reference.

2.3   RESULTS AND DISCUSSION

2.3.1   MORPHOLOGIES OF AL COATINGS

SEM micrographs of the Al-coated surfaces of the films were taken to observe the morphology of the coatings on the films. The Al coating on the surface of the films were broken by stretching the polymer phase to observe the thickness of Al coatings.

2.3.1.1   MORPHOLOGY OF COMMERCIAL FILM 1

The commercial polypropylene films were coated with a layer of Al by physical vapor deposition. The coating thickness was determined from SEM pictures of commercial film 1 as can be seen in Figure 2.1(a) as 32 nm for commercial film 1. The coating had the shape of the polymer layer. There were parallel lines on the surface of the polymer produced during processing of the polypropylene film in continuous film machinery.

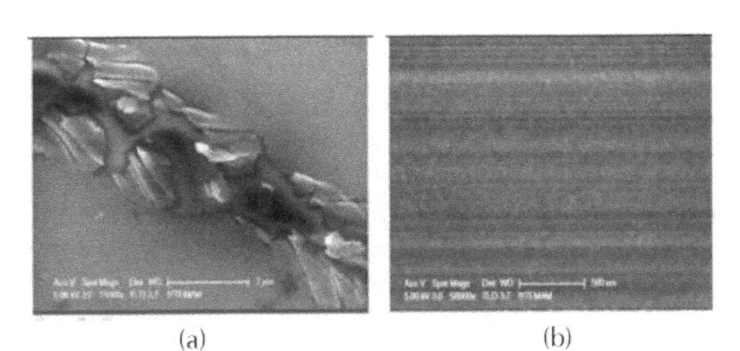

(a)   (b)

FIGURE 2.1   SEM pictures of al-coated surface of commercial film 1: (a) the al surface fractured to observe coating thickness and (b) the surface as produced.

2.3.1.2   MORPHOLOGY OF COMMERCIAL FILM 2

In Figure 2.2, SEM micrograph of commercial film 2 is shown. Figure 2.2(a) is fractured Al coating on the surface. Coating thickness is determined as 185 nm from Figure 2.2(a). Figure 2.2(b) is Al-coated polypropylene surface. The coating was made up of Al particles condensed on the surface of the polypropylene phase. There were uncoated regions on the surface.

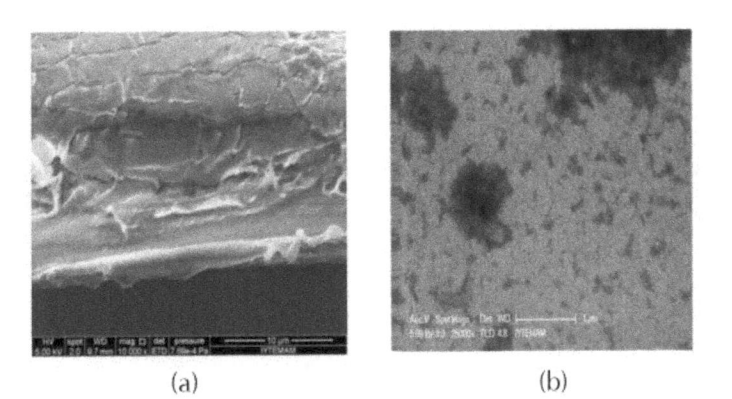

(a)   (b)

FIGURE 2.2   SEM Pictures from BOPP metalized commercial film 2(a). Fractured Al surface (b). Coated polypropylene.

2.3.1.3 MORPHOLOGY OF MAGNETRON SPUTTERED FILMS

The polypropylene films were at 50 µm thickness and their one surface was covered with 98–131 nm thick Al layer with magnetron sputtering [18]. In Figure 2.3, SEM micrographs of magnetron sputtered film are shown. In the Figure 2.3(a) there is fractured Al coating on the surface of polypropylene. Figure 2.6(b). is the top view of the Al layer. From Figure 2.3(a), the coating thickness is determined to be 185 nm, close to the value reported previously [18]. The top view of the Al layer seen in Figure 2.3(b) indicated that the surface was covered by Al particles having nearly 30 nm size. The surface was not very smooth. There are cracks on the surface of the brittle Al coating.

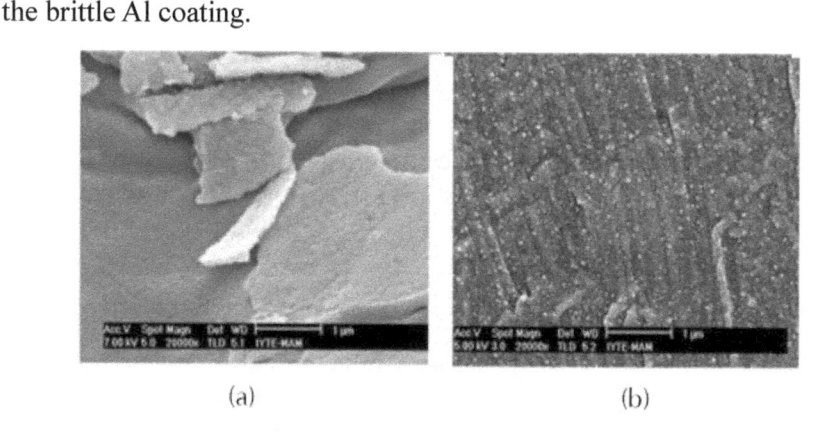

(a)   (b)

FIGURE 2.3 SEM pictures of a. fractured Al coating on the cast polypropylene film (b). Al coated surface of the polypropylene film.

(a)   (b)

FIGURE 2.4 The photographs of (a) Chocolate packing (b) Milk cover.

2.3.1.4   MORPHOLOGY OF MILK AND CHOCOLATE PACKAGES

The chocolate packing had an outer layer coated with aluminum. The inner layer was milk cover that had the function of opening the packing when it was pulled. The inner side of the cover that was in contact with the milk was covered with a polymer layer.

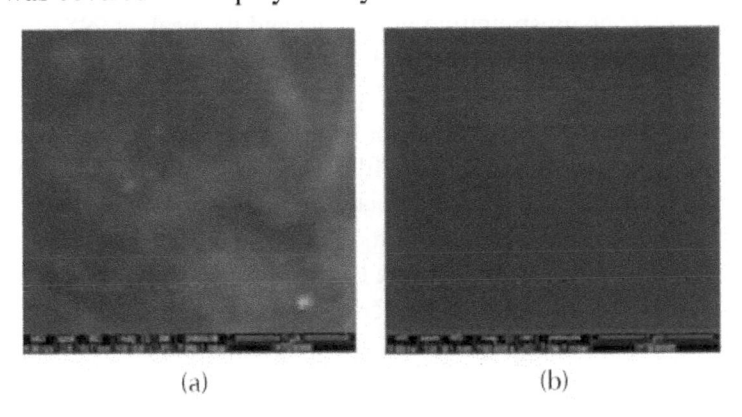

(a)   (b)

FIGURE 2.5   The SEM micrographs Al coated surface of a. the chocolate packing (b). Milk cover.

Photographs of chocolate packing and milk cover are shown in Figure 2.4. The SEM micrographs of their Al-coated surfaces are shown in Figure 2.5. They have an even Al coating as seen in Figure 2.5.

2.3.2   TRANSMISSION SPECTRA OF AL-COATED POLYPROPYLENE FILMS

2.3.2.1   COMMERCIAL AL-COATED POLYPROPYLENE FILMS

The transmission infrared spectra of commercial films 1 and 2 were obtained by placing the films in the path of the infrared light. Spectrum 1 and spectrum 2 in Figure 2.6 belonged to commercial film 1 and film 2, respectively. The baseline absorbance values were close to 1 for commercial film 1 and were close to 4 for commercial film 2. This indicated that 1/10000 and 1/10 of the input infrared light was transmitted from the samples. The Al coating on the film acted as a barrier to infrared light. The higher coating thickness of commercial film 2 (185 nm) than that of commercial film 1 (30 nm) resulted higher level of shielding from infrared radiation. The film 2 had better infrared light shielding efficiency than the film 1.

The peaks observed at 3000–2800 cm^{-1} belonged to C–H asymmetric and symmetric stretching, at 1450 cm^{-1} C–H bending, at 1350 cm^{-1} C–H deformation bending[17]. The peak at 973 cm^{-1} belongs to amorphous CH_3 rocking and C–C chain stretching vibrations and the peak at 998 cm^{-1} belongs to crystalline CH_3 rocking, CH_2 wagging, and CH bending vibrations[20].

The magnetron sputtered film's polymer layer and Al layer were 50 μm and 185 nm, respectively. This film was nearly three times thicker than commercial film 1 and commercial film 2. The absorbance values of this film in transmission mode were too high, since both the Al layer and polypropylene layer were thicker. The FTIR transmission spectrum of this thick film coated by magnetron sputtering had very high absorbance values, indicating it was also a good shield for infrared radiation.

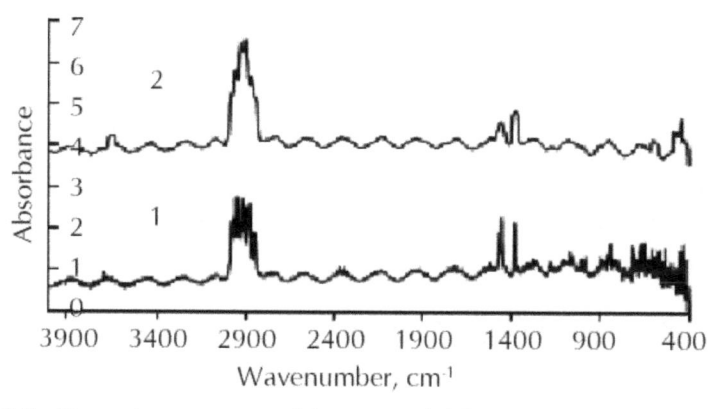

FIGURE 2.6   Transmittance spectra of the commercial films 1 and 2.

2.3.3 SPECULAR REFLECTANCE OF SPECTRA OF THE FILMS

2.3.3.1 COMMERCIAL AL-COATED POLYPROPYLENE FILM1

Figure 2.6 shows the specular reflectance spectra of Al-coated and uncoated surfaces of the film 1. Al-coated surface reflected 100 percent of the infrared rays since the absorbance values were very close to zero.

Uncoated surface had the characteristic spectrum of polypropylene. However, there were other peaks observed called fringes due to reflection of light from both surfaces of the thin film (Figure 2.7).

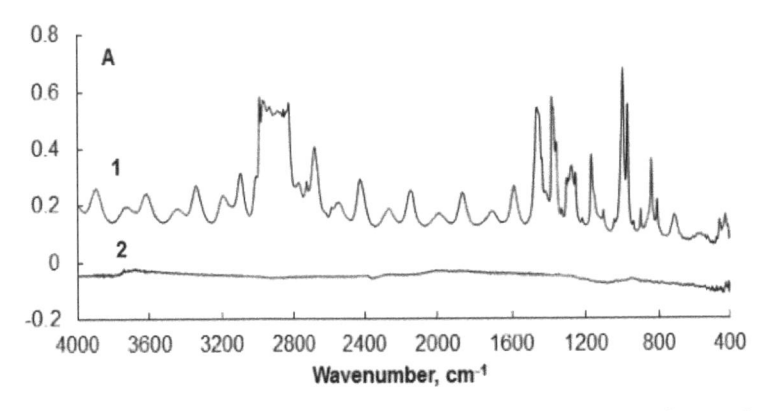

FIGURE 2.7 Specular reflectance of spectra of the film 1, (1) Polypropylene surface, (2) Al-coated surface of the film 1.

2.3.3.2 THICKNESS OF COMMERCIAL FILM 1

To count the number of fringes, select the starting and ending points both as minima or maxima of the spectrum and then count the number of opposing minima or maxima. In other words, if we select minima values for starting and ending points in the spectrum, then select maxima points to count the number of fringes.

2.3.3.3 SPECULAR REFLECTANCE SPECTRA OF COMMERCIAL AL-COATED POLYPROPYLENE FILM 2

The specular reflectance spectra of the surfaces of the commercial film 2 are seen in Figure 2.8; curve 1 shows the specular reflectance spectrum of the polypropylene side of the film and curve 2 shows the specular reflection spectrum of the Al side of the film. Although the polypropylene surface has the spectrum of polypropylene, the Al-coated surface reflected the infrared rays. The absorbance values close to zero indicated that the light was not absorbed but reflected by the Al surface.

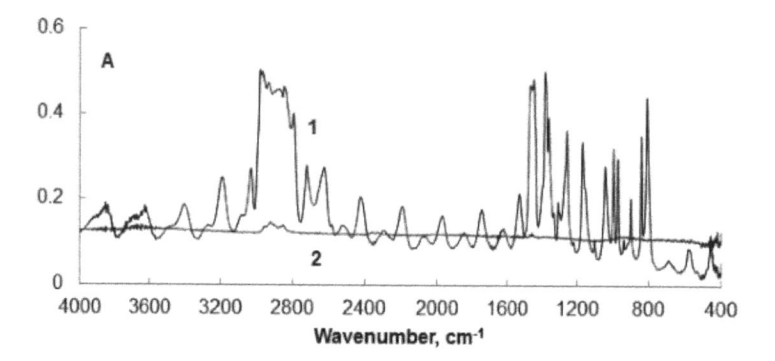

FIGURE 2.8   Specular reflectance spectra of the film 2: 1, polypropylene surface and 2, Al-coated surface of the film 1.

2.3.3.4   THICKNESS OF COMMERCIAL FILM 2

There were six refraction fringes observed in specular reflection spectrum of commercial film 2 in Figure 2.5. Using Eq. (2.1) and inserting values of six fringes between 1600 cm^{-1} and 2650 cm^{-1} and the film thickness was calculated to be 19 μm.

2.3.3.5   SPECULAR REFLECTANCE SPECTRA OF MAGNETRON SPUTTERED FILMS

The specular reflection spectra of uncoated and coated surfaces of polypropylene film prepared by Ozmihci et al.,[5] can be seen in Figure 2.9. Both surfaces showed the

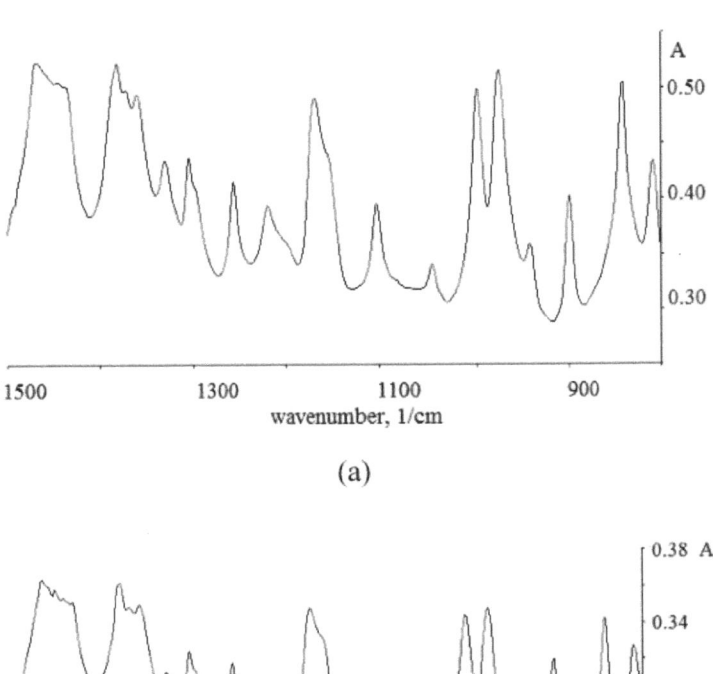

(a)

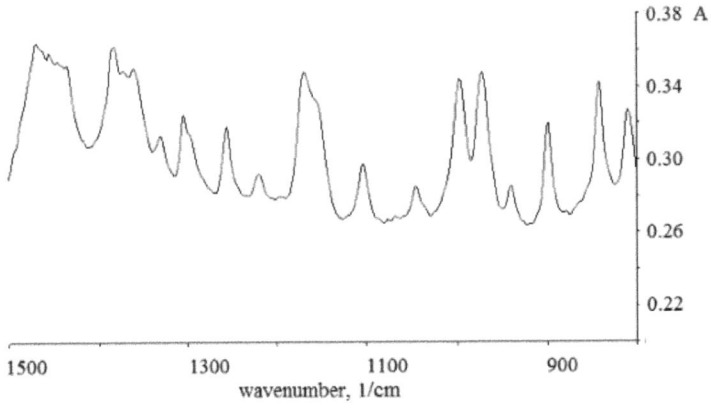

(b)

FIGURE 2.9 Specular reflection spectra of (a) Uncoated surface and (b) Al-coated surface of cast film.

Characteristic spectrum of polypropylene. This indicated that there were uncoated polypropylene regions on the coated surface. However the Al-coated surface had lower absorbance values at all wave numbers due to reflection of infrared rays from its surface. However, the reflection extent was not as high as the reflection extent of the commercial films. The Al surface of this film was not a reflecting surface like commercial films.

2.3.3.6 SPECULAR REFLECTANCE SPECTRA OF MILK PACKAGE

Milk cardboard container has an Al lid to open it. The specular reflection spectra of uncoated and coated surfaces of milk package can be seen in Figure 2.10. Al coating at the upper surface has lower absorbance value than polymer at the lower surface. Thus, Al coating reflected the infrared light and caused light protection of milk from heat and light conditions. The lower surface is less bright. Both surfaces appear to be covered with different polymer layers. The shining surface had peaks at 2920 cm^{-1}, 2860 cm^{-1}, 1745 cm^{-1}, 1645 cm^{-1}, 1282 cm^{-1}, 1070 cm^{-1}, 997 cm^{-1}, and 869 cm^{-1}. There are CH_2 streching vibration peaks at 2920 and 2860 cm^{-1}, C = O peaks at 1745 cm^{-1}, C = C peak at 1645 cm^{-1}. The other surface had peaks at 2951 cm^{-1}, 2885 cm^{-1}, 1734 cm^{-1}, 1454 cm^{-1}, 1296 cm^{-1}, and 723 cm^{-1}, respectively. 2951 cm^{-} and 2885 cm^{-1} peak belonged to CH_2 stretching vibrations and 1745 cm^{-1} peak belonged to C = O stretching vibration. Further characterizations are needed to determine which polymers were coated on the surfaces.

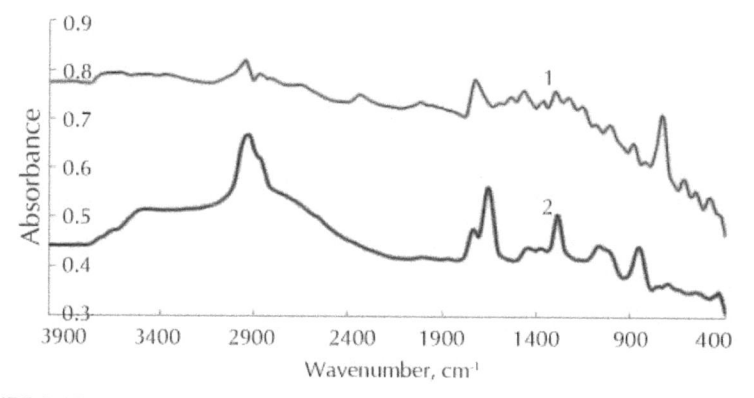

FIGURE 2.10 Specular reflection of milk package: 1, Mat surface and 2, shining surface.

2.3.4 SPECULAR REFLECTANCE SPECTRA OF CHOCOLATE PACKING

The chocolate packing showed that Al-coated surface had lower absorbance values than uncoated surface indicating that infrared light was reflected by the Al coating as can be seen in Figure 2.11.

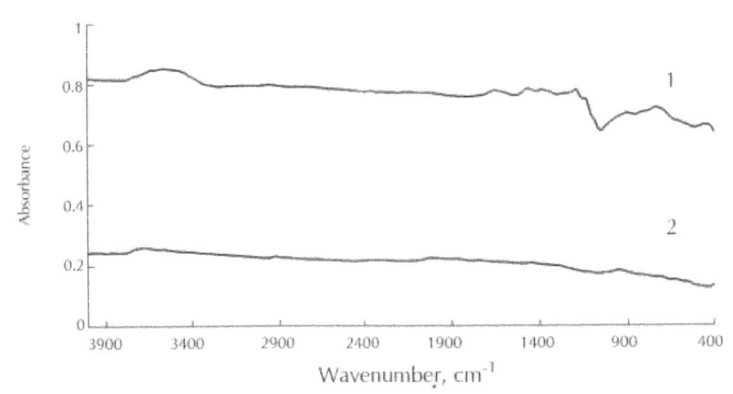

FIGURE 2.11 Specular reflection spectra of chocolate package: 1, Polymer surface and 2, Al-coated surface.

2.3.5 COMPARISON OF ABSORBANCE VALUES IN SPECULAR REFLECTANCE SPECTRA OF PACKING MATERIALS

Comparison of five different films absorbance values are given in Table 2.1. Baseline absorbance values at 920 cm^{-1} and maximum absorbance values due to polypropylene band at 1450 cm^{-1} are reported in the table. The baseline absorbance values of the uncoated and coated surfaces at 920 cm^{-1} are close to each other for all five films. The highest baseline value was observed for magnetron sputtered films. The lowest baseline absorbance was observed for commercial film 1. The absorbance values of the uncoated polypropylene surface of the five films at 1450 cm^{-1} are close to each other in the range of 0.50–0.77. This peak is due to bending vibration of the CH_2 groups in polypropylene phase. FTIR light was reflected by the Al surfaces of the films. The highest reflection and the lowest absorbance value of –0.08 was observed for commercial film 2. Commercial film 1 also had a low absorbance value of 0.11 and high level of reflection. The magnetron sputtered films reflection was not as high as the commercial films, since its absorbance value at 1450 cm^{-1} was 0.36. The Al coating on this film did not cover the whole surface, and there were polypropylene phase exposed to infrared light. Chocolate packing materials' Al-coated surface also had low absorbance value, 0.20, indicating that it also reflected strogly the infrared light. The milk cover Al-coated surface even higher absorbance value 0.44 than other films.

TABLE 2.1 Absorbance values of specular reflectance spectra of uncoated and Al-coated surfaces of packing materials

Type of film	Absorbance at 1450 cm^{-1}		Absorbance at 920 cm^{-1}	
	Uncoated Surface	Al-Coated Surface	Uncoated Surface	Al-Coated Surface
Commercial film 1	0.50	0.11	0.12	0.11
Commercial film 2	0.66	−0.08	−0.03	−0.07
Magnetron sputtered film	0.60	0.36	0.29	0.28
Chocolate packing	0.77	0.20	0.69	0.18
Milk cover	0.75	0.44	0.63	0.38

2.4 CONCLUSIONS

Infrared spectroscopy is an efficient tool for measuring the thickness of thin polymer films and their ability to absorb or reflect infrared lights. The thicknesses of the commercial films coated with aluminum were determined to be 16 and 19 μm for film 1 and 2, respectively. The aluminum-coated surface of the commercial films had the ability to reflect the infrared rays that strike. They can be used efficiently as infrared light shields for the materials inside their packing. The Al coatings obtained by chemical vapor deposition were 32 nm and 185 nm for commercial film 1 and 2, respectively. The coatings were more perfect than the coating obtained by magnetron sputtered cast film.

The Al coating of the cast film was more brittle, and there were uncoated polypropylene regions, and the infrared light was only partially filtered. Thus, it was a less-efficient infrared shield when compared with commercial films. The magnetron sputtering method needs more investigation for optimum results. The chocolate and milk-packing materials' Al-coated surfaces reflected the infrared light better than their other surfaces. They were also good infrared light protectors. However, milk-packing material Al surface was also coated with another polymer layer either for esthetic or safety reasons or since it was used in contact with milk.

ACKNOWLEDGMENTS

The authors thank The POLİNAS and POLİBAK companies and the authors of Ref. [19] for providing for providing commercial aluminum-coated films and magnetron sputtered film, respectively.

KEYWORDS

- **Aluminum-coated polymer films**
- **commercial films**
- **quantum-size effect**

REFERENCES

1. Aihong G.; Xuejiao T.; Sujuan Z.; *Key Eng. Mater.* **2011**, *474–476*, 195–199.
2. Dombrowsky, L. A.; *Rev. Gen. Therm.* **1998**, *37*, 925–933.
3. King, D. E.; Drewry, D. G.; Sample, J. L.; Clemons, D. E.; Caruso, K. S.; Potocki, K. A.; Eng, D. A.; Mehoke, D. S.; Mattix, M. P.; Thomas, M. E.; Nagle, D. C.; *Int. J. Appl. Ceram. Technol.* **2009**, *6(3)*, 355–361.
4. Shahidi, S.; Ghoranneviss, M.; Moazzenchi, B.; Anvari, E.; Rashidi, A.; *Surf. Coat. Technol.* **2007**, *201*, 5646–5650.
5. Takano, I.; Inoue, N.; Matsui, K.; Kokubu, S.; Sasase, M.; Isobe, S.; *Surf. Coat. Technol.* **1994**, *66*, 509–513.
6. O'Hare, L.; A., Leadley, S.; Parbhoo, B.; *Surf. Interface Anal.* **2002**, *33*, 335–342.
7. Greer, J.; Street R.; A.; *Acta Mater.* **2007**, *55*, 6345–6349.
8. Fortunato, E.; Nunes, P.; Marques, A.; Costa, D.; Aguas, H.; Ferreira, I.; Costa, M. E. V.; Godinho, M. H.; Almeida, P. L.; Borges, J. P.; Martins, R.; *Surf. Coat. Technol.* **2002**, 151–152, 247–251.
9. Yanaka, M.; Henry, B., M.; Roberts, A. P.; Grovenor, C., R., M., Briggs, G., A., D., Sutton, A., P.; Miyamoto, T.; Tsukahara, Y.; Takeda, N.; Chjater, R., J., *Thin Solid Films.* **2001**, *397*, 176–185.
10. Bichler, C. H.; Kerbstadt, T.; Langowski, H. C.; Moosheimer, U. *Surf. Coat. Technol.* **1999**, *112*, 373–378.
11. Bichler, C. H.; Langowski, H. C.; Moosheimer, U.; Seifert, B.; *J. Adhes. Sci. Technol.* **1997**, *11(2)*, 233–246.
12. Moosheimer, U.; Bichler, C. H.; *Surf. Coat. Technol.* **1999**, *116*, 812–819.
13. Oishi, T.; Goto, M.; Pihosh, Y.; Kasahara, A.; Tosa, M.; *Appl. Polym. Sci.* **2005**, *241*, 223–226.
14. Qi-Jia He; Ai-Min Zhang; Ling-Hong Guo; *Polym. Plast. Technol. Eng.* **2004**, *43(3)* 951–961.

15. Qui J.; Liu L. H.; Hsu P. F.; *Appl. Surf. Sci.* **2005,** *111*, 1912–1920.
16. Rahmatollahpur R.; Tahidi T.; Jamshidi-Chaleh K. *J. Mat. Sci.* **2010,** *45*, **1937.**
17. Ozmihci, F.; Balkose, D.; Ulku, S.; J. *Appl. Polym. Sci.* **2001,** *82*, 2913–2921.
18. Balkose, D.; Oguz, K.; Ozyuzer, L.; Tari, S.; Arkis, E.; Omurlu, F.O.; *J. Appl. Polym. Sci.* **2012,** *120*, 1671–1678.
19. http://www.piketech.com/skin/fashion_mosaic_blue/application-pdfs /Calculating Thickness- Free Standing Films-by FTIR.pdf, **2012.**
20. Luongo, JP.; *Infrared study of polypropylene. J. App. Polym. Sci.* **1960,** *9*, 302–309.

CHAPTER 3

GENERALIZATION OF FUELS SWELLING DATA BY MEANS OF LINEAR FREE ENERGY PRINCIPLE

ROMAN MAKITRA, HALYNA MIDYANA, LILIYA BAZYLYAK, and
OLENA PALCHYKOVA

CONTENTS

3.1 INTRODUCTION

An action of the organic solvents on solid fuels, namely black and brown coal or peat is under the active investigation for a long time. This is caused by two reasons; specifically, this is one among successful methods of the solid fuels structure study with the aim of their technological application for obtaining of the so-called montan wax, or it is used for obtaining of the low-molecular liquid extracts that can be next transformed into synthetic liquid fuel via the hydrogenation. It is necessary to add that the interaction of coal with the solvents is taken as a principle of the coals liquefaction under their transformation into the liquid fuel.

3.2 METHODOLOGY

The first stage of the interaction into the systems *coal solvent* is their swelling and expansion as a result of the molecules of liquid invasion into the pores and into the structure of coal. Depending on the nature of solvent and also on coal itself, its volume can be multiplied many times; correspondingly, the increment of weight for the investigated sample can achieve 100 and more percentage.

A review of the early works concerning coal swelling is presented in Refs. [1–2]. It can be noted that still in 50 years, the coals swelling process was considered as the first stage of their extraction, and it was connected with their physical–chemical interaction with the solvents [3]. The process itself was explained by two reasons: first is caused by the adsorption and absorption of liquid in pores, and accordingly to the second one the swelling process is connected with the change of the cohesion energy of the solid and liquid phases of a system. Sanada and coworkers [4–6] considered that the coal represents by itself the natural three–dimensional polymer, and in accordance with the Flory—Huggin's theory, the change of a free energy under the coal swelling represents by itself the conditional sum of the polymer and the solvent mixing energies; this free energy change is determined first of all by the difference in the Hildebrand's solubility parameters δ of both components:

$$\Delta G = [ln(1-\varnothing_2) + \chi\varnothing^2 + \varnothing_2],\qquad(3.1)$$

where $\varnothing_2$ is a volumetrical part of the netting structure (of polymer) into the swollen system; and the parameter χ indicating the interaction of polymer and the solvent is equal to

$$\chi = \frac{\beta + (\delta_1 - \delta_2)^2 V_1}{RT} \tag{3.2}$$

where δ_1 and δ_2 are the Hildebrand's parameters of the solubility for the solvent and for the polymer, and they are equal to $[(\Delta H_{vap} - RT)/V_{mol}]^{1/2}$, respectively; the empirical term β is the correction factor taking into account the number of the branching points into the structure of polymer.

After the insignificant transformations, it can be obtained the Flory–Renner's equation. Such equation helps to calculate the sizes of the polymer link between the cross bonds M_c:

$$M_c = \frac{\rho_2 V_1 \varnothing_2^{1/3}}{[-\ln(1-\varnothing_2) - \chi \varnothing_2^2 - \varnothing_2]}, \tag{3.3}$$

where ρ_2 is the density of a polymer into solution; V_1 is molar volume of the solvent.

Coefficient χ is determined empirically for each solvent and, of course, from the concentration dependence of the osmotic pressure taking into account a series of the assumptions.

In the work of Sanada [5] and subsequent works of other authors, the swelling degree in the volumetric parts Q is represented as a function on the δ of solvents. These data in the most cases form the parabolic, a so-called *bell-like* curve with a maximum for the solvent, δ_2 of which in accordance with the theory of the regular solutions is equal or is near to δ_1 of a polymer (coal). In reality, in Ref. [5], it was already maintained that only the approximated dependencies are obtained for coal since a number of experimental data concerning to Q are visibly take one's leaved from the generalizing curve. Under the extraction of vitrain from the Yubary field (the content of carbon consists of 85.2 percent) using the solvents into Soxhlet's apparatus it was determined in Ref. [4], that the maximal yields of the extract are observed at their molar volume about 10 cm³/mole (ethylendiamine, dimethylformamide, cyclohexanone 24–26 percent, acetophenone 35.6 percent, pyridine 33.2 percent); this fact authors explain by the influence of the value of cohesion energy.

However, there are a plenty of exclusions, for example, for butanol V_{mol} = 9.5, but the yield of an extract in this case is only 0.8 per cent. The explanation of this deviation as a result of the solvent association, which caused by the presence of the hydrogen bond, seems unconvincing, since under the experiments conditions (the extraction proceeds into Soxhlet's apparatus, and, therefore under the boiling temperature) such association will be insignificant. It was discovered in Ref. [6] that the value M_c for Japanese coal with the carbon content less than 80 per cent is unreal low—only 10 (!), next this value, is sharply increased and achieves the maximum M_c = 175.

Authors starting from the following positions at the explanation of this fact: firs, experimental determinations were carried out in pyridine, in which the specific interactions can take place and, second, this deviation can be caused by the mistakes at the determination of χ coefficient. It is necessary to notify that although it is hardly to estimate the verisimilitude of determined in such a way molecular weights of structural links of a coal between the points of cross bonds, however, in a case of the synthetic polymers in a same way determined masses of links visible do not agree with the values obtained in accordance with others methods. The same approach was discussed especially in detail in the work of Kirov and coauthors [7] on example of swelling (and extraction) of three kinds of bituminous Australian coal. These authors confirmed the main observations of Sanada, namely the swelling degree Q increases from ~1.4 in hydrocarbons to ~2 in pyridine (δ–11.0) and again is decreased to ~ 1.5 in alcohols. Calculated on this basis value δ of coal is increased steadily at the increase of the carbon content from 70 percent till 87 percent and in a case of more metamorphosed coal is sharply decreased again.

Data concerning to the extraction of *Greta* coal are evidence of the maximal yield of extract (more than 20%) under its treatment with ethylendiamine and dimethylformamide (δ–11.5); however, authors admit a fact that this is a consequence of the specific interactions, since in alcohols from the δ by the same order, the yield of the extract is only 1–2 percent. But for all that authors concluded that although the swelling degree of coal is not directly connected with the molecular characteristics of the absorbed liquids, however the determining factor is their parameter of the solubility despite the fact that at the detailed consideration of the dependencies Q = $f(\delta)$ (or $f(\delta^2)$), there are a number of deviations (as same as in Ref. [5]) from the ideal curve for many solvents are observed.

Despite the indicated lacks, the above described approach is applied in later works concerning the coal swelling for the results interpretation. It is necessary to indicate a plenty of investigations devoted to the swelling studies of coal No 6 from the Illinois State (standard in the US coal for the carbon–chemical investigations) [8–10]. General conclusions are in good agreement with the results of Ref. [5, 7]. Comparison of the swelling degree for different coal in some solvents depending on the content of carbon has been done in Ref. [11]. Similar investigation for Kansk–Achynsk Siberia coal was carried out in Ref. [12, 13]. In both cases, as same as in Ref. [7], it was proved that the dependence of swelling degree of coal on the carbon content in it.

It is logical to assume the influence of the specific interactions during the swelling process because the values of the parameters of coal solubility δ^2, which are determining accordingly to the Flory–Renner's equation, are differed. It depends on fact that the data for all solvents are taken into account in calculations, or such calculations are performed with the exclusion of results for the solvents that can be acceptors of the hydrogen bonds (amines, ketones). Different results have been obtained also at the application of other calculations methods, especially of the Van–Krevelen's method [14].

It is notified in Ref. [15] that the swelling of some coal does not agree with the thesis of the regular solutions theory; that is why, to calculate the parameter χ for them is impossible. Authors explain this fact by the presence of oxygen atoms in the investigated coal. But also the molecular mass of separate sections (clusters) between the points of crossing for methylated or acetylated samples is equal only to 300–600 in accordance with the calculations (i.e., unreal).It is necessary to notify that the critical analysis of the Flory theory application for the determination of molecular mass and the crossing density of the coal structure has been done in the Painter's works [16]. Authors assert that the possible formation of the hydrogen bonds between the hydroxy groups of the low–metamorphized coal and the basic solvents plays a significant role here; this is why even the introduction of a lot of the empirical amendments into the calculations leads to the obtaining of the understated values of molecular masses of clusters.

Taking into account the above-mentioned lacks, many authors concluded that the theory of regular solutions is insufficient for adequate description of the coals swelling process in different solvents (and also for their extraction) since such theory does not take into account the possible specific solvation of the active structures of coal and first of all its hetero-atoms [17], especially by the formation of the hydrogen bonds. To take into account the possible acid–base interactions, it was proposed by Mar-zec and coauthors [18, 19] to determine the swelling degree as a function of donor number of DN solvents or as a function of their donor and accep-tor numbers disparity accordingly to Gutmann. However, corresponding analysis of data concerning the swelling of the Silessian bituminous coal showed the following: although between the swelling degree Q and DN, the visible symbasis exists and the deviations from the straight line are less than for the function $Q = f(\delta)$, however, it is complicated to confirm about the quantitative description of the process. The same conclusion about only qualitative character of such dependence has been done by authors in Ref. [13] on example of swelling the brown Kansk–Achynsk coal and some kinds of the Donbas coal.

Above-mentioned facts and disagreements lead to the conclusions in Ref. [20] that the sorption of solvents by coal is very complicated process covering also the changes under the action of a solvent into the coal struc-ture and other possible nonequilibrium phenomena. That is why an appli-cation for coal of the theories developed for the description of thermody-namically equilibrium process of the simple synthetic polymers swelling is unwarranted first of all due to neglect by existing chemical (specific) solvation interactions. As it was confirmed in many investigations, the swelling Flory–Huggins's model based on the theory of regular solutions is not sufficiently consistent with the real experimental data. It is caused by a range of simplifying assumptions putted into the base of this model and, first of all, by the presence of full isoentalpic mixing (solution) of two phases that is in disagreement with the reality—even in a case of the polymers that do not contain the donor–acceptor groups into the structure a swelling and solution processes are accompanied with a great enthalpy effect; it is known, that even nonspecific solvation is often accompanied by the changes of free energy and enthalpy of the system. Also, isoentalpy will be not remained in a case of the possible donor–acceptor (acid–base) interaction, which is often observed in a case of the synthetic polymers with the content of heteroatoms (polyurethane, nitryle rubbers) and is ob-served in a case of coal as a result of the presence in it of such groups as –OH, –COOH, tertiary atom of nitrogen and etc.

The principle of the linearity of free energies (LFEs) is applied in chem-istry of solutions over 30 years for quantitative description of the solvents influence on the behavior of dissolved substances (spectral characteristics, constants of the reaction rate). In accordance with this principle, general change of free energy of the system consists of the separate interindependent terms and first of all consists of the nonspecific and specific solvation effects and also needed energy for the formation of cavity in the structure of liquid phase with the aim of allocation the exterior molecule introducing there. And only full sum of these all possible energetic effects gives the final (equilibrium) energy of the system [21]:

$$\Delta G = \sum \Delta g_i \qquad (3.4)$$

By taking this into account, the constants of the reaction rates are determined via the equilibrium constants of the activated reactive complex formation, and the last in part depends on the solvation processes; Koppel and Palm [22] proposed the following equation to determine the influence of medium properties on the reaction rates of the processes proceeding in it:

$$\lg k = a_0 + \frac{a_1(n^2 - 1)}{(n^2 + 2)} + \frac{a_2(\varepsilon - 1)}{(2\varepsilon + 1)} + a_3 B + a_4 E_T \qquad (3.5)$$

The above equation takes into account the influence of the polarization $f(n^2)$ and polarity $f(\varepsilon)$ of the solvents determining their ability to nonspecific solvation and also their basicities B, and electrophilicity accordingly to Reichardt, E_T characterizing their ability to introduce into acid–base interactions (specific solvation). Appropriateness of this equation for the generalization of experimental data of the dependencies of reactions rates (and also spectral characteristics of dissolved substances) on physical–chemical characteristics of the solvents has been proved by a number of hundred examples.

For the processes of the phase equilibria, it was proposed by us to add to the Koppel–Palm equation the fifth term, which takes into account the density of the energy of solvents cohesion proportional to the squared Hildebrand's solubility parameter δ^2. Due to this fact, necessary energy for the formation of cavity for the allocation of the molecule introducing into liquid phase is taking into account:

$$\lg K = a_0 + a_1 f(n^2) + a_2 f(\varepsilon) + a_3 B + a_4 E_T + a_5 \delta^2 \qquad (3.6)$$

Modified equation turned out to be effective for determining the solvents' influence on the equilibrium of such processes as solubility in different media of gases or solids, the distribution of substances between two phases, and resembling equilibrium processes. Therefore, it will be logical to try to use Eq. (3.6) for the swelling processes. As a matter of fact, it turned out that with the use of the equation, it is possible to determine the quantitative connection between the properties of the solvents and the equilibrium swelling degree for a number of polymers, and also for a coal [23–25].

To achieve the satisfactorily high values of the coefficients of multiple correlations R, it is necessary to exclude from the calculations the data for some quantity (3–5) of the solvents. It is hard to explain this. Besides, the model of the interactions into the system was not quite clear. In a case when the solvation processes are energetically advantageous ($\Delta G < 0$), the role of δ^2 factor remains unclear. Such factor characterizes the energy needed for the cavity formation into the structure of the liquid; at the same time, unlike the evaporation process, under the swelling of substances into liquid, the following process takes place: liquid solvent penetrates into the structure of solid polymeric phase mostly as the whole.

At the beginning of the ninetieth century, the works of Aminabhavi appeared [26]. These worked concerned the polymeric membranes swelling into organic solvents and to the diffusion rate D of the liquids into their structure in which these values were considered as dependencies from the molar volume V_{mol} of the liquids. Generalizations obtained in Ref. [26] are rather unsatisfactory–approximately linear dependencies lgQ or lgD on V_{mol} are observed only in the homologic ranges or in the case of similar solvents. But the approach itself must be considered as logical: it is clear that the bigger the sizes of the introducing molecule, the more difficult it will be to penetrate into the structure of polymer including the adsorbent interstice. Low generalization ability of the dependencies presented in the work [26] can be explained by fact that they do not take into account the solvation effects that promote to liquids penetration. That is why Eq. (3.6) has been expanded by us by the additional term taking into account the influence of the molar volume V_{mol} of the solvents [27]:

$$\lg Q = a_0 + a_1 f(n^2) + a_2 f(\varepsilon) + a_3 B + a_4 E_T + a_5 \delta^2 + a_6 V_M \qquad (3.7)$$

Such equation under the stipulation that Q is represented not in the volumetric parts accordingly to the Flory–Huggins's model but in accordance with the interpretation of equilibrium processes in the chemical thermodynamics as a moles of the solvent absorbed by one gram or by one cm^3 of polymer turned out to be effective under the generalization of data for swelling degree of different synthetic polymers, for example, polyethylene, in different organic solvents depending on their physical–chemical parameters [28]. That is s why it was necessary to check the possibility of Eq. (3.7) application for the generalization of data concerning coal swelling since this equation takes into account the following: (i) All the most important possible energetic effects caused by possible donor–acceptor interaction of active groups of the coal with noninert solvents including the formation of hydrogen bonds; (ii) the effects of nonspecific coal solvation with solvents, which are caused by a presence in it the cyclic aromatic structures as a result of which the visible influence of the ability of some solvents for the polarization can be expected; (iii) the endothermic effects as a result of steric complications of the solvents penetration (V_M); and (iv) destruction of the liquid phase structure (δ^2).

The meaningfulness of the parameter V_M was confirmed in Ref. [29–30], where it was shown that the quantity of the solvent's moles absorbed by the coal weight unit is decreased symbasis to the V_M of the solvent; at the same time, in Ref. [31] it was shown that under the coals swelling in homologous series of the solvents by the same basicity (amines, alcohols, etc.), the value lgQ is linearly decreased till the number of Carbon's atoms, that is, to their V_M. In addition, the values of the correction parameter χ from Eq. (3.1) can be connected with the physical–chemical characteristics of the penetrants via Eq. (3.7) (see Ref. [32]).

Data concerning the swelling of the most popular coal (namely, coal Illinois № 6) have been considered as the main object of our investigations. This coal is the standard object for the carbon–chemical investigations in the United States. These data were already analyzed earlier in Refs. [23–25] with the aim of their generalization according to Eq. (3.6), but the results obtained were unsatisfactory. Evidently, this was caused by two factors: (i) the influence of the molecules sizes of the solvents penetrating into the coal structure (their molar volume) was not taken into account in these works; (ii) starting values of the swelling degree Q were given accordingly to original works [9–11] in ml (sometimes in g) of the solvent absorbed by 1 ml or 1 g of coal since the Flory–Huggins's model has been used by authors (this model uses the volumes of two liquids that are mutually mixed). If to consider the coal swelling process (and generally polymers swelling processes) as thermodynamically equilibrium process then the free energy change at the penetration.

In this chapter, we have checked the efficiency of the factors mentioned above, considering studying the swelling process in organic solvents. Illinois coals are low–metamorphized, bituminous and contain 20–31 percent of volatile substances, are characterized by ash content 8–12 percent and sulfur content 4–7.8 percent. Investigated in Ref. [8–10] sample was characterized by following composition: C 79.8; H 5.11; N 1.8; S_{org} 2.0 and O 11.2 percent. The samples were pulverized and extracted from the soluble components with pyridine [8] and were washed by water, and after vacuum drying, they were saturated by solvent's steams until their full saturation at room temperatures in the closed vessels. Authors give the ratio of the weights for swelled samples respectively to the starting W; they were recalculated in the quantities of solvent moles absorbed by 1 g of coal S_M (Table 1).

TABLE 1 Experimental [8] and calculated in accordance with the two-parametric Eq. (3.8) values of the swelling degree S_M of the coal Illinois № 6

№	Solvent	Experiments			Calculations		
		W	$S_M 10^3$	lgS_M	lgS_M	ΔlgS_M	Δ %
1	Pyridine	1.87	11.00	−1.959	−1.958	−0.001	0.1
2	Dioxane	1.64	7.264	−2.139	−2.143	0.005	0.2
3	CH_2Cl_2	1.60	7.064	−2.151	−2.179	0.028	1.3
4	Clorobenzene	1.48	4.264	−2.370	−2.364	−0.006	0.3
5	Isopropanol	1.45	7.489	−2.126	−2.136	0.010	0.5
6	Toluene	1.41	4.450	−2.352	−2.364	0.012	0.5
7	Ethanol	1.40	8.682	−2.061	−2.069	−0.008	0.4
8	Benzene	1.38	4.864	−2.313	−2.283	0.030	1.3
9	Acetonitrile	1.34	8.283	−2.082	−2.056	0.026	1.2
10	Cyclohexene*	1.11	1.307	−2.884	–	–	–

Note: *Data excluded from the final calculations

Generalization of studied data for 10 solvents in accordance with the fifth-parameter Eq. (3.6) leads to the expression with unsatisfactory low value of correlation coefficient $R = 0.81$ [23]. Exclusion from the consideration of the most uncoordinated data for dioxane gives the possibility to obtain the fifth-parameter equation with low, but acceptable degree of connection $R = 0.941$. At the same time, consideration of the molar volume factor and the change of weight parts on the molar ones essentially improve the correlation for all 10 studied solvents $R = 0.940$, and after the exclusion the data concerning to cyclohexene for the rest 9 solvents we obtain the equation with high connection degree $R = 0.996$ [33].

There are only two decisive parameters here *the basicity* that assists to the swelling process and *the molar volume*, which opposites to this process, taking into account the needed energy and the negotiation of the cohesion forces have only insignificant influence and the exclusion of this parameter from the calculations practically does not worsen the equation:

$$\lg S_M = -1.96 + (0.665 \pm 0.074)10^{-3} B - (4.12 \pm 0.58)10^{-3} V_M;$$

$$R = 0,984$$

and

$$S = 0,030 \tag{3.8}$$

Obtained equations have greater predicted ability comparatively to the fifth-parameter equation obtained in Ref. [23], which does not take into account the factor of the molar volume.

The influence of the factor of molar volume is confirmed by fact that between $\lg S_M$ and V_M the neatly marked symbasis is observed, namely with increasing of V_M the value $\lg S_M$ is decreased. The action of other decisive factor solvents basicity is opposite, in other words, with the basicity increasing the symbate increasing of $\lg S_M$ is observed. Evidently, this is caused by the specific solvation of acid centers, which are in the macromolecule of coal and, first of all, of hydroxy groups. Their presence can be assumed taking into account a great number of the oxygen in the Illinois coal. Comparison of these two oppositely directed dependencies leads to the conclusion, that they are mutually compensated. Although these dependencies are only symbate, but the algebraic sum of the influence of these two factors is practically linearly connected with the respective values lgS_M [34]. The third factor is the density of the cohesion energy, which is proportional to needed energy for separation of absorbed molecules from the structure of liquid phase; this factor respectively also decreases the swelling value. However, the influence of this value is insignificant; this fact is confirmed by negligible decreasing of the R value at its exclusion only from 0.991 to 0.984. The possible processes of nonspecific and electrophilic solvation practically do not impact on the value S_M.

In Ref. [10], authors also have been studied the swelling process of the coal Illinois № 6 in the liquid phase. Swelling degree S_V has been studied volumetrically as the ratio of volumes of swelling sample to the starting one. Unlike Ref. [9], the process was investigated in a range of amines including the primary ones, which were able to form hydrogen bonds and also alcohols. At the generalization of these data in accordance with the fifth-parameter equation without taking into account of V_M for the all 17 solvents, the equation was obtained with the low value $R = 0.861$; in order to obtain the satisfactory correlation, it is necessary to exclude from the consideration the data for the methyl and dimethylanilines [25].

TABLE 2   Experimental [10] and calculated in accordance with the Eq. (3.10) values of swelling degree for the coal *Illinois № 6*

№	Solvent	Experiments			Calculations	
		S_V	$S_M 10^3$	lgS_M	lgS_M	ΔlgS_M
1	2-Picoline	2.76	17.84	−1.749	−1.794	0.045
2	Pyridine	2.75	21.72	−1.663	−1.808	0.145
3	Butylamine	2.64	16.57	−1.781	−1.941	0.160
4	Propylamine	2.45	17.62	−1.754	−1.655	−0.099
5	Aniline	1.99	10.86	−1.964	−2.205	0.241
6	2-Hexanon	1.98	8.120	−2.090	−2.312	0.221
7	Methylaniline	1.44	4.052	−2.392	−2.037	−0.355

TABLE 2 *(Continued)*

№	Solvent	Experiments			Calculations	
		S_V	$S_M 10^3$	lgS_M	lgS_M	ΔlgS_M
8	Propanol	1.36	4.820	−2.317	−2.240	−0.077
9	Ethanol	1.34	5.824	−2.235	−2.194	−0.041
10	Butanol	1.34	3.715	−2.430	−2.242	−0.188
11	Methanol	1.23	5.690	−2.245	−2.204	−0.041
12	Dimethylaniline*	1.10	0.789	−3.103	—	—
13	Isopropanole*	1.06	0.784	−3.106	—	—
14	Toluene	1.06	0.562	−3.250	−3.311	0.061
15	p-Xylene	1.06	0.487	−3.312	−3.326	0.013
16	m-Xylene	1.05	0.407	−3.390	−3.293	−0.097
17	Benzene	1.04	0.447	−3.350	−3.362	0.012

Note: *Data excluded from the calculations

In Table 3.2, the starting values of S_V from Ref. [10] are presented and calculated on the basis quantities of solvent's moles absorbed by 1 g of coal–S_M and lgS_M. In this case for 17 solvents, it was obtained by means of the six-parametric Eq. (3.7) a relatively low value of R (only 0.884) too, but after the exclusion of data for isopropanol and dimethylaniline, we will obtain Eq. (3.9) with a high correlation degree:

$$\lg S_M = -2.91 + (0.454 \pm 1.40) f(n^2) + (5.73 \pm 1.22) f(\varepsilon) + (1.43 \pm 0.37) 10^{-3}$$
$$B - (29.9 \pm 22.0) 10^{-3} E_T - (0.722 \pm 0.947) \delta^2 - (6.52 \pm 4.54) 10^{-3} V_M$$

$$N = 15 \quad R = 0.981 \quad S = 0.160 \tag{3.9}$$

and after the exclusion of insignificant factors

$$\lg S_M = -4,34 + (3.42 \pm 0.69) f(\varepsilon) + (2.02 \pm 0.27) 10^{-3} B - (0.86 \pm 2.55) 10^{-3} V_{mol}$$

$$R = 0.968 \text{ and } S = 0.172 \hspace{3cm} (3.10)$$

In this case, *the basicity* and *the molar volume* of the solvents are decisive factors too, the influence of which is oppositely directed. An appearance of the polarity as significant factor is connected with the specific selection of high polar solvents (alcohols, amines).

Accordingly to Ref. [9], the swelling process of the Illinois coal № 6 has been carried out principally under other conditions, namely the samples were previously extracted with pyridine, dried coal was stranded till the full saturation with vapors at 100°C in closed metallic ampoules (with the exception of phenol, investigating temperature of which is 182°C). Authors presented the results of investigations as the ratio of swelling W (in percentages); that is, the ratio of weights of swelling sample after 1 h to the dried sample. These data have been previously generalized in Ref. [24]. Low value R for the all 12 solvents equal to 0.876 after the exclusion from the consideration data concerning to the phenol and tetrahydrophurane is increased till 0.972. Essentially, better results were obtained by taken into account the molar volume factor and after recalculation of W in S_M.

The data concerning to W taken from Ref. [9] and calculated on their basis swelling values in moles S_M and lgS_M are presented in Table 3.3; the generalization of these data in accordance with the sixth-parameter Eq. (3.7) leads to higher degree of relationship $R = 0.909$, and the exclusion from the consideration of one solvent (butylamine) gives the possibility to obtain the equation with satisfactory degree of relationship $R = 0.974$; an additional exclusion of the dimethylformamide gives Eq. (3.11) with $R = 0.991$:

$$lg Q = -2,61 + (3,50 \pm 0,82) f(n^2) + (2.30 \pm 0.46) f(\varepsilon) - (0.33 \pm 0.14)10^{-3}$$
$$B - (2.37 \pm 6.8)10^{-3} E_T + (0.70 \pm 0.24)10^{-3} \delta^2 - (1.5 \pm 2.1)10^{-3} V_M$$

$$N = 10, \ R = 0.991 \text{ and } S = 0.055 \hspace{2cm} (3.11)$$

and after the exclusion of insignificant factors

$$lg Q = -2,51 + (2,66 \pm 1,20) f(n^2) + (1.80 \pm 0.60) f(\varepsilon) - (7.4 \pm 5.0)10^{-3} E_T - (2.8 \pm 2.9)10^{-3} V_M$$

$$R = 0.964 \text{ and } S = 0.086 \hspace{3cm} (3.12)$$

TABLE 3 Experimental [9] and calculated in accordance with the Eq. (3.12) values of swelling ratio of soluble part of coal for the coal Illinois № 6

№	Solvent	Experiments			Calculations	
		W	$S_M 10^3$	lgS_M	lgS_M	ΔlgS_M
1	Dimethylformamide	6.2	60.54	−1.218	—	—
2	N-Methylpyrrolidone	5.7	37.17	−1.430	−1.503	0.073
3	Dimethylsulphoxide	5.5	49.19	−1.308	−1.407	0.099
4	Ethylendiamine	4.6	33.24	−1.478	−1.466	−0.012
5	Aniline	4.6	34.13	−1.467	−1.513	0.047
6	Butylamine*	3.8	29.82	−1.525	—	—
7	Pyridine	3.7	28.67	−1.543	−1.475	−0.068
8	Phenol	3.4	22.49	−1.648	−1.512	−0.136
9	Pipyridine	3.0	19.26	−1.543	−1.475	−0.068
10	Tetrahydrofuran	2.8	22.97	−1.639	−1.615	−0.024
11	Toluene	2.6	16.65	−1.779	−1.874	0.095
12	Hexane	1.6	6.902	−2.161	−2.134	−0.027

Note: *Data excluded from the calculations

With the increase in molar volume of the solvents, the coal swelling degree gets decreased; the same is an effect of the ability to electrophilic solvation. Unlike both previous cases, the positive influence of the solvents basicity (namely their ability to form the donor–acceptor bonds with acid groups of the coal) here is insignificant evidently as a consequence of especial influence of the conditions of experiment carrying out. Under higher temperatures, the hydrogen bonds get easily decomposed. At the same time, the possible positive influence of the factors of nonspecific solvation $f(n^2)$ and $f(\varepsilon)$ is observed. Calculated values lgS_M and their discrepancy with the experiment ΔlgS_M are presented for comparison in Table 3.3.

Both the equilibrium swelling degree and the kinetics of this process depend on the character of the solvent. In Ref. [10], it has been studied that the swelling rate of the coal Illinois № 6 volumetrically in different solvents; on the starting stages, it is ordered to the pseudo first-order reactions kinetics as is observed in the case of polymers swelling too. It helps in determining the respective constants rate of the process k, which are presented in Table 3.4. In Ref. [25], we have generalized these data for 24 solvents with the use of the fifth-parameter Eq. (3.6). For the all maximal sequence of the data, the value of correlation multiple coefficient R was very low and equal to 0.694; and only after the exclusion from the calculation, the data for five solvents (that is practically 20 percent) the satisfactory value of $R = 0.957$ can be obtained. In addition, taking into account the influence of molar volume, that is transition to sixth-parameter equation, gives the possibility to obtain the expression with $R = 0.883$. And in order to obtain the satisfactory correlation, it was enough to exclude from the calculations data for only two solvents, namely 2–hexanone (methylbutyl ketone) and triethylamine:

$$\lg k = 2.20 - (2.55 \pm 3.84) f(n^2) + (1.08 \pm 4.26) f(\varepsilon) + (4.17 \pm 1.12)10^{-3} B - (71.7 \pm 62.5)10^{-3} E_T - (0.92 \pm 2.45)\delta^2 - (42.8 \pm 9.1)10^{-3} V_M$$

$$N = 22, \quad R = 0.959 \quad \text{and} \quad S = 0.448 \qquad (3.13)$$

The equation terms characterizing the influence of non–specific solvation and also cohesion energy have a great standard deviations which are more than the absolute values of the coefficients and that is why are evidently insignificant. Checking the value R decreasing at the exclusion of these terms confirmed this assumption and helped to obtain the equation with lesser quantity of significant terms. This equation also adequately characterizes the influence of the solvents properties on the rate of their penetration into the coal structure; besides, the decisive factor in this case as same as in a case of swelling value is the influence of molar volume of the solvents, increasing of which leads to the process rate decreasing.

$$\lg k = 1.12 + (4.85 \pm 0.52)10^{-3} B - (66.0 \pm 19.5)10^{-3} E_T - (42.1 \pm 6.2)10^{-3} V_M$$

$$R = 0.957 \quad \text{and} \quad S = 0.418 \qquad (3.14)$$

Significant factor as same as in a case of the swelling degree is the solvents basicity. With the solvents basicity increasing, the process rate is also increased. The less essential is a role the solvents ability to electrophilic solvation; although this factor increases the process rate but it exclusion from the consideration decreases R till 0.928. The values lgk calculated in accordance with the Eq. (3.14) are represented in Table 3.4.

TABLE 3.4   Experimental [10] and calculated accordingly to Eq. (3.14) values of the logarithms of the constants rate of the coals *Illinois № 6* swelling

№	Solvent	Experiments		Calculations	
		$k10^5, s^{-1}$	lgk	lgk	Δlgk
1	Propylamine	1167.0	−1.933	−1.679	−0.254
2	Butylamine	614.0	−2.212	−2.974	0.763
3	Pyridine	316.7	−2.499	−2.655	0.155
4	2-Picoline	126.7	−2.897	−3.124	0.226
5	2-Hexanone*	125.0	−2.903	—	—
6	Methanol	53.30	−3.273	−3.183	−0.091
7	Ethanol	21.30	−3.672	−3.624	−0.048
8	Aniline	20.0	−3.699	−3.962	0.263
9	Propanol	10.90	−3.963	−4.290	0.328
10	Butanol	3.84	−4.416	−4.926	0.511
11	Isopropanol	2.54	−4.595	−4.152	−0.443
12	Methylaniline	2.47	−4.607	−4.063	−0.545
13	Butanole-2	1.32	−4.879	−4.694	−0.185
14	Toluene	0.90	−5.046	−5.334	0.288
15	Isobutanol	0.833	−5.079	−4.859	−0.220
16	Dimethylaniline	0.59	−5.229	−4.578	−0.651
17	Benzene	0.45	−5.347	−4.675	−0.671
18	Pentanol	0.376	−5.425	−5.510	0.085
19	*p*-Xylene	0.375	−5.426	−5.924	0.498
20	*m*-Xylene	0.225	−5.648	−5.919	0.271
21	*o*-Xylene	0.118	−5.928	−5.891	−0.037
22	Ethyl benzene	0.113	−5.947	−6.032	0.086
23	Cumene	0.009	−7.046	−6.716	−0.330
24	Triethylamine*	0.00038	−8.420	—	—

Note: *Data excluded from the calculations

Decisive role of the V_{M} factor during the adsorption process of the solvents by coal is in agreement with the determined in Ref. [25] proportionality for the alcohols between lgk and steric factor E_{s} of the Hammet–Taft's equation.

Therefore, the swelling characteristics of the Illinois coal are determined by total influence of molar volume of liquids and their ability to specific solvation. The same conclusion has been done by authors Refs. [35–36] explaining the adsorption growing by increasing the donor number of the solvents via the formation of hydrogen bond by OH–groups of coal. But these authors have not done respective quantitative generalization giving the possibility on the basis of the linearity of free energies principle adequately to connect the properties of the liquids with their ability to interact with a coal; it was confirmed that the approaches based on the theory of regular solutions equitable only at the consideration of the swelling process in the inert (so-called low basic) solvents, mainly of low–polar.

Correctness of the sixth-parameter Eq. (3.7) and its simplified forms for the generalization of the swelling data was proved for other coals including the Donbas coal [37] at the parameters B and V_{M}, for lignites from the *Rawhile* and *Big Brown* mines (USA) and also for the *Silesian* coal (Poland). In Ref. [18] presented data concerning to the bituminous *Silesian* coal swelling in 21 solvents. It can be assumed, that such coal is situated between the cannel coals and gas coal (such assumption has been done on the basis of the *Silesian* coal characteristics, namely (percent per daf): $C = 80.7$; $H = 5.6$; $N = 1.9$ and $S = 0.9$; a content of the volatile substances consists of 39.5 percent). The swelling degree Q has been determined volumetrically at 25°C starting from the increase of the volume of coal under the solvent excess till the equilibrium achievement (2–7 weeks). Appropriating values of Q are represented in Table 3.5.

TABLE 5　Experimental [8] and calculated values of the bituminous *Silesian* coal swelling degree

№	Solvent	Experiments			Calculations		
		Q	$S_{M}10^{3}$	lgS_{M}	lgS_{M}	ΔlgS_{M}	$\Delta\%$
1	Benzene*	1.00	0.00	–	–	–	–
2	Nitrobenzene	1.10	0.98	–3.009	–3.003	0.006	0.2

TABLE 5   *(Continued)*

№	Solvent	Experiments			Calculations		
		Q	$S_M 10^3$	lgS_M	lgS_M	ΔlgS_M	$\Delta \%$
3	Diethyl ether*	1.15	1.44	−2.842	−	−	−
4	Isopropanol*	1.14	1.83	−2.738	−	−	−
5	1,4-Dioxane	1.16	1.88	−2.726	−2.704	−0.022	0.8
6	Ethyl acetate	1.26	2.66	−2.575	−2.481	−0.094	3.6
7	Acetonitrile	1.15	2.86	−2.544	−2.499	−0.045	1.8
8	n-Propanol	1.23	3.08	−2.511	−2.316	−0.196	7.8
9	Nitromethane*	1.18	3.36	−2.474	−	−	−
10	Methyl acetate	1.32	3.99	−2.399	−2.528	0.128	5.4
11	Acetone	1.30	4.08	−2.389	−2.366	−0.023	1.0
12	Ethanol	1.25	4.28	−2.369	−2.316	−0.052	2.2
13	Methanol	1.19	4.70	−2.328	−2.402	0.075	3.2
14	Methyl ethyl ketone	1.49	5.47	−2.262	−2.256	−0.006	0.2
15	Dimethoxy-ethane	1.60	5.75	−2.240	−2.327	0.087	3.9
16	Tetrahydrofuran	1.59	7.28	−2.138	−2.138	0.000	0.0
17	Dimethyl formamide	1.69	8.92	−2.050	−2.118	0.069	3.3
18	Pyridine	2.08	13.40	−1.873	−1.711	−0.162	8.6
19	N-Methylpir-rolidone	2.38	14.28	−1.845	−1.912	0.067	3.6
20	Dimethyl sulfoxide	2.04	14.66	−1.834	−2.014	0.180	9.8
21	Ethylenedi-amine*	2.08	16.17	−1.791	−	−	−

Note: *Data Excluded from the Final Calculations

Authors in Ref. [8] consider that inherently of the presented coal swelling process is the donor–acceptor interaction of the coals' acidic groups with the basic ones of the solvents. However, really they determined only the symbasis between the swelling degree Q and donor number of the solvent DN; at that, incomprehensible sharp bend is observed at $DN = 12-13$ on this dependence. In a case of hydrocarbons, ethers, alcohols it is observed only the slow increasing of Q at the DN increase, while for the carbonyl compounds of amines and dimethyl sulfide (DMSO) the inclination of line is essentially steeper [8]. The same peculiarity is observed also for the S_M. At the same time, the dependence between Q and δ was not determined.

In Ref. [24], we had a go to generalize the results presented in Ref. [8] via the fifth-parametric Eq. (3.6) without taking into account a factor of the molar volume. The obtained value R for 21 solvents was unsatisfactory (0.887); and in order to achieve the acceptable value of $R = 0.958$, it was necessary to exclude from the consideration the data for three solvents.

In accordance with the approach developed in Ref. [29], we had a go to generalize the data from Ref. [8] by means of Eq. (3.1), that is, taking into account of influence of the molar volume of the solvents V_M; the logarithm of the moles of solvent absorbed by 1 sm^3 of coal lgS_M determined based on the W (see Table 3.5) was used as the correlated value. However, in a case for 20 solvents (data for benzene are excluded from the consideration since for this solvent W and, respectively, S_M are equal to zero) the correlation is also unsatisfactory: $R = 0.818$; sequential exclusion of data for diethyl ether (№ 3), isopropanol (№ 4), nitromethane (№ 9) and ethylenediamine (№ 21) increases the value R till 0.877; 0.911; 0.934 and 0.969. Thus, for the 16 solvents, the sixth-parametric Eq. (3.15) was obtained, which with the acceptable accuracy connects the swelling degree with the physical–chemical characteristics of the solvents:

$$\lg S_M = -2.72 - (6.07 \pm 1.45)f(n) + (2.26 \pm 0.55)f(\varepsilon) + (3.53 \pm 0.33)10^{-3}B - (0,130 \pm 0,063)E_T + n$$
$$+ (1.17 \pm 0.55)10^{-3}\delta^2 - 9.92 \pm 3.36)10^{-3}V_M$$

$$N = 16, \quad R = 0.969 \quad \text{and} \quad S = 0.104 \tag{3.15}$$

By–turn exclusion of the separate terms of an equation denotes on the lesser significance of the terms with E_T (the R of the corresponding fifth-parametric equation is equal to 0.954) and with the δ^2 (R is equal to 0.951).

Thereby, the swelling process of the presented coal can be adequately de‐scribed with the use of the fourth‐parametric equation:

$$\lg S_M = -3,47 - (4,11 \pm 1,27)f(n) + (1.82 \pm 0.51)f(\varepsilon) + (3.50 \pm 0.36)10^{-3}B - (6.66 \pm 2.45)10^{-3}V_M$$

$$(3.16)$$

$$N = 16, \quad R = 0.951 \text{ and } S = 0.117$$

At the same time, the equations which do not take into account the fac‐tors of the non‐specific solvation give the essentially poor results:

$$\lg S_M = n3.02 + 3.0 \cdot 10^{-3}B - 3.85 \cdot 10^{-3}V_M \, ; R = 0.872; S = 0.171.$$

And the three‐parametric equations with taken into account of $f(n)$ or $f(\varepsilon)$ give the value or $R \sim 0,9$.

In Table 3.5, the values of lgS_M are presented, which were determined in accordance with the above presented equation. As can be seen, the de‐viations more than ± 0.117 are observed only for the propanol, pyridine and *DMSO*. Essentially, more deviations were observed also for the ex‐cluded from the calculations solvents. In a case of the isopropanol such disagreement of the calculations and experiment was caused probably by the steric difficulties, and in a case of the ethylene diamine—by the extreme high basicity. It's interesting, that in a case of the benzene, in which the swelling practically does not proceed (S_M is equal to zero), the calculation accordingly to the Eq. (3.16) gives the values $lgS_M = -3,49$, i. e., it is dif‐fered only insignificantly from zero by the value S_M.

It is necessary to stop on a role of the separate terms into Eq. (3.16). Between lgS_M and V_M the symbate decrease of the swelling degree with the liquid molar volume increasing is observed (as same as in the previous cases). In accordance with the earlier developed principles, it very im‐portant is also the contribution of the acid—base interaction—the swelling degree is distinctly increased at the value of B of solvents (see Figure 3.1), though the correlation between these values is only medium ($R = 0.752$ for the all points).

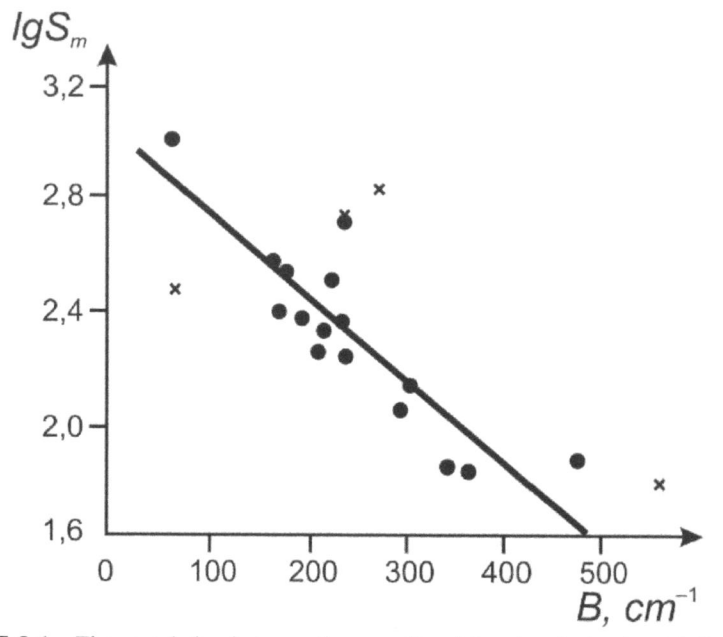

FIGURE 3.1 The correlation between the quantity of the absorbed solvent lgS_M by the bituminous silesian coal and the basicity B; ' − data excluded from the calculation.

After the exclusion of the most deviating solvents, the value R becomes equal to 0.870. Significantly is also the influence of the nonspecific solvation. After the exclusion of one among terms $f(n)$ or $f(\varepsilon)$ the R is decreased till 0.9, and in a case of the exclusion of both together terms this value is decreased till 0.872. But if to apply Eq. (3.7) to the coal extraction data, then the factor of molar volume V_M is insignificant, and the connection between quantities of extracted substance (in g/mole of the solvent) and physical–chemical characteristics can be satisfactorily described by fifth parameter Eq. (3.6) or by its simplified forms; in this case possible acid–base interaction is the decisive factor, that is factor B [38–40]

This confirmation is in good agreement with the above–said: bigger molecules harder introduce into the coal structure and after equilibrium state their size does not play the role. Let us notify, that the same approach has the positive results at the data generalization concerning to the solubility of the synthetic low–molecular coal analogous diphenylolpropane in 20 solvents [41]. This approach is also applicable for the generalization of data concerning to the coal extraction under sub–critical conditions, but the role of the specific solvation is also insignificant, evidently as a result of its suppression at high temperatures.

Therefore, it was discovered the lack of fit the description of the coal swelling process with the use of one–parametric dependencies including those dependencies based on the theory of regular solutions on the solubility parameter of liquids. It was shown, that the quantitative connection between the swelling degree of coal and physical–mechanical properties of the solvents is achieved only on the basis of principle of the linearity of free energy under condition of taking into account the all-solvation process. The basicity of the solvents and their molar volume are the factors determining the swelling degree for low-metamorphized coal.

KEYWORDS

- **Coal swelling**
- **Fuels swelling**
- **Linear free energy principle**

REFERENCE

1. Kibler, M. V.; The solvents action on coals In Book Chemistry of Solid Fuels **1**, **1951**, 145–267 *(in Russian)*.
2. Keller, D. V.; Smith, C. D.; Spontaneous fracture of coal. *Fuel*. **1976**, *55(4)*, 272–280.
3. Kröger, K.; Die steinnohleextraction. *Erdöl und Kohle*. **1956**, *9(7)*, 441–446.
4. Sanada, Y.; Honda, H.; Solvent extraction of coal bull. *Chem. Soc. Japan*. 1962, *35(8)*, 1358–1360.
5. Sanada, Y.; Honda, H.; Equilibrium swelling of coals in various solvents. *Fuel*. 1966, *45(4)*, 451–456.
6. Sanada, Y.; Honda, H.; Swelling equilibrium of coals by pyridine. *Fuel*. 1966, *45(4)*, 295–300.
7. Kirov, N. Y; O'Shea, J. N; Sergeant, G. D.; The determination of solubility parameters of coal. *Fuel*. 1967, *47*, 415–424.
8. Green, T. K; Kovac, J.; Larsen, J. W. A; Rapid and convenient method for measuring the swelling of coals. *Fuel*. 63(7), 1984, 935–938.
9. Mayo, F. R.; Zevely, J. S.; Pavelka, L. A.; Extractions and reactions of coals below 100 °C. *Fuel*. 1988, *67(5)*, 595–599.
10. Aida, T.; Fuku, K.; Fujii, M. et al.; Steric requirements for the solvent swelling of Illinois № 6. *Coal Energy Fuels*. 1991, *5(6)*, 74–83.

11. Nelson, J. F; Mahajant, O. T.; Walker, P. L.; Measurement of swelling of coals in organic liquids. *Fuel.* 1980, *59(12)*, 831–837.
12. Skrypchenko, G. B.; Khrennikova, O. V.; Rybakov, S. I.; *Khimiya Tviordogo Topliva.* 1987, *5*, 23–28 *(in Russian)*.
13. Osipov, A. M.; Bojko, Z. V.; *Khimiya Tviordogo Topliva*, 1987, *3*, 15–18 *(in Russian)*.
14. Van Krevelen, W.; Chemical structure and properties of coal. *Fuel.* 1965, *44(4)*, 229–242.
15. Larsen, J. W.; Shawyer, S.; Solvent swelling studies. *Energy Fuels.* 1990, *4(1)*, 72–74.
16. Painter, P. C.; Graf, J.; Coleman, M. H.; Coal solubility and swelling. *Parts* 120133. *Energy Fuels.* 1990, *4(4)*, 379–397.
17. Weyrich, O. R.; Larsen, J. W.; Thermodynamics of hydrogen bonding in coal–derived liquids. *Fuel.* 1983, *62(8)*, 976–977.
18. Marzec, A.; Kisielow, W.; Mechanism of swelling and extraction and coal structure. *Fuel.* 1983, *62(8)*, 977–979.
19. Szeliga, J.; Marzec, A.; Swelling of coal in relations to solvent electron–donor numbers. *Fuel.* 1983, *63(10)*, 1229–1231.
20. Hsieh, S. T.; Duda, J. L.; Probing coal structure with organic vapor sorption. *Fuel.* **1987**, *66(2)*, 170–178.
21. Mayer, U.; Eine Semiempirische Gleichung zur Beschreibung des Lösungs–Mitteleinflusses auf Statik und Kinetik Chemischer Reaktionen. *Th.* 1, 2. *Monutsh. Chemie.* 1978, *109(H. 2)*, 421–433, 775–790.
22. Koppel, I. A.; Palm, V. A.; The influence of the solvent on organic reactivity. In: *Advances in Linear Free Energy Relationships.* Ed. Chapman, N. B. & Shorter, J.; London, New York: Plenum Press; 1972, 203–281.
23. Makitra, R. G.; Pyrig, Ya. M.; *Khimiya Tviordogo Topliva.* 1988, *6*, 41–45 *(in Russian)*.
24. Makitra, R. G.; Pyrig, Ya. M.; *Khimiya Tviordogo Topliva.* 1992, 6, 11–20 *(in Russian)*.
25. Makitra, R. G.; Pyrig, Ya. M.; Vasiutyn, Ya. M.; *Khimiya Tviordogo Topliva.* 1995, *3*, 3–13 *(in Russian)*.
26. Aminabhavi, T. M.; Harogopadd, S. B.; Khinnavar, R. S. et al.; Rubber solvent interactions. *Rev. Macromol. Chem. Phys.* 1991, *C 31(4)*, 433–497.
27. Makitra, R. G.; Turovsky, A. A.; Zaikov, G. E.; Correlation analysis in chemistry of solutions. Eds. *VSP, Utrecht–Boston*; 2004, 324 p.
28. Makitra, R. G.; Pyrig, Ya. M.; Zaglad'ko, E. A.; *Plastics.* 2001, *3*, 23–27 *(in Russian)*.
29. Makitra, R. G.; Midyana, G. G.; Palchykova, E. Ya.; Bryk, D. V.; *Solid Fuel Chem.* 2010, *44(6)*, 407.
30. Makitra, R. G.; Prystansky, R. Ye.; *Khimiya Tviordogo Topliva.* 2003, *4*, 24 *(in Russian)*.
31. Makitra, R. G.; Bryk, D. V; *Solid Fuel Chem.* 2010, *44(3)*, 164.
32. Makitra, R. G.; Midyana, G. G.; Palchykova, E. Ya.; *Solid Fuel Chem.* 2011, *45(6)*, 430.
33. Makitra, R.; Polyuzhyn, I.; Prystansky, R.; Smyrnova, O.; Rogovyk, V.; Zaglad'ko, O.; Application of free energies linearity principles for sorption and penetration of the organic substances. *Works of the Scientific Society Named After Shevchenko.* 2003, *10*, 152–163 *(in Ukrainian)*.

34. Makitra, R. G.; Prystansky, R. Ye.; *Khimiya Tviordogo Topliva*, 2001, *5*, 316 *(in Russian)*.
35. Larsen, J. W.; Green, T. K.; Kovac, J.; J. Org. Chem., *50(10)*, 1985, 4729–4735.
36. Hall, P. G.; Marsh, H.; Thomas, K. M.; Solvent induced swelling of coals to study macromolecular structure. *Fuel.* 1988, *67(6)*, 863–866.
37. Makitra, R. G.; Prystansky, R. Ye.; *Khimiya Tviordogo Topliva*. 2003, *4*, 24–36 *(in Russian)*.
38. Vasiutyn, Ya. M.; Makitra, R. G.; Pyrig, Ya. M.; Turovsky, A. A.; *Khimiya Tviordogo Topliva*, 1994, 4, 66–73 *(in Russian)*.
39. Makitra, R. G.; Pyrig, Ya. M.; *Khimiya Tviordogo Topliva*. 1991, 1, 67–70 *(in Russian)*.
40. Makitra, R. G.; Pyrig, Ya. M.; *Khimiya Tviordogo Topliva*. 1993, *3*, 14–18 *(in Russian)*.
41. Makitra, R. G.; Bryk, S. D.; Palchykova, O. Ya.; An investigation of interaction of low–metamorphized coal with organic solvents (on example of diphenilolpropane). *Geol. Geochem. Combust. Miner.* 2003, *3–4*, 126–130 *(in Ukrainian)*.

CHAPTER 4

TRENDS ON NEW BIODEGRADABLE BLENDS ON THE BASIS OF COPOLYMERS 3-HYDROXYBUTYRATE WITH HYDROXYVALERATE AND SEGMENTED POLYETHERURETHANE

SVETLANA G. KARPOVA, SERGEI M. LOMAKIN, ANATOLII A. POPOV, and ALEKSEI A. IORDANSKII

CONTENTS

4.1 INTRODUCTION

The new biodegradable blends based on the combination of synthetic and natural polymers are alternative option for the use of individual polymers. As a result of composite production, the emergence of essentially new exploitation characteristics is expected, which are not inherent to the original polymer components. Biodegradable systems are widely used in innovative technologies for the drug delivery, in tissue engineering, and prosthetic vascular stenting, in a contemporary design of environmentally friendly barrier materials in packaging [1–3]. When creating such systems, segmented polyetherurethanes (conventional SPEU on the basis of multifunctional isocyanates) may be of particular interest. Because of the unique combination between mechanical properties and biocompatibility, they are widely and successfully applied in various fields of biomedicine as constructional and functional materials [4, 5]. However, the low rate of SPEU biodegradation, which is a positive factor in the case of a long-term functioning, limits considerably their use during short-term exploitation. Regulation of SPEU lifetime can be achieved by blending it with a variety of biodegradable biopolymers, such as poly(α-hydroxyacid)s or poly(β-hydroxyalkanoate)s and the common representatives of the latter as poly(3-hydroxybutyrate) (PHB) and its copolymer with 3-hydroxyvalerate (PHBV) (Refs [6, 7]).

PHB, along with valuable properties, has certain disadvantages, in particular, high cost and fragility. For overcoming such disadvantages, either copolymers PHAs or their blends with other natural polymers, in particular, with chitosan [8] are often used.

PHB–SPEU and PHBV–SPEU polymer systems can be applied as the basis of promising new composites, in particular, for use in cardiosurgery, as well as for a scaffold design. By varying the blend content, and, consequently, affecting the morphology of the systems, we can obtain composition with different physical and chemical characteristics such as permeability, solubility in water and drugs, controlled release, and reduced rate of degradation. The comprehensive description of PHBV–SPEU system requires the assessment of a combination of structural and dynamic features that allow a more complete evaluation of its structural evolution at relatively short-term times that precede its hydrolytic decomposition as a result of exposure to water at elevated temperatures.

Blends comprising the copolymer of 3-hydroxybutyrate (~ 95mol percent) and 3-hydroxyvalerat (~5 mol percent) PHBV (Tianan Co, Ningbo, China), M_w = 2.4 × 105, M_n = 1.5 × 105, ρ=1.25 g/cm3) and SPEU on the basis of MDI medical grade (Elastogran, Basf Co., Germany), M_w = 2.29×105, M_n = 5.3 × 104, ρ = 0.97 g/cm3 were studied. Mixing weight ratio of PHBV/SPEU ranged in the following sequence: 100:0, 60:40, 50:50, 40:60, and 0:100 percent, respectively. The films were obtained in a solvent mixture of tetrahydrofuran and dioxane. In addition, by the DSC method, the thermophysical characteristics (melting points and endothermal fusion) were studied. DSC measurements were performed with a microcalorimeter DSC 204 F1 Netzsch Co in an inert atmosphere of argon at a heating rate of 10°/min. Analysis of the mixtures was carried out by dynamic (ESR probe analysis and isotope deuterium D–H exchange) and structural (WAXS and DSC) methods. Segmental mobility was studied by paramagnetic probe by determining the correlation time τ, characterizing the rotational mobility of the nitroxyl probe TEMPO in accordance with known method [9, 10].

X-ray analysis of the films was carried out using transmission technique with a diffractometer Bruker Advance D8 (Cu Kα). IR spectra of deuterated films were recorded on a spectrophotometer Bruker IFS 48 with a Fourier transform at a resolution of 2 cm^{-1}.

4.2   X-RAY AND DSC STUDIES

In this study, it was shown by X-ray technique (WAXS) that the original samples of PHBV are characterized by high crystallinity. In their diffraction patterns, at least five reflections of the orthorhombic lattice with characteristics a = 5.74 Å, b = 13.24 Å and c = 5.98 Å were determined that well conform to the earlier studies [11, 12]. In Figure 4.1 the WAXS diffractograms are presented as a set of curves belonging to initial PHBV (1), PHBV treated by water at 40°C (2), PHBV/SPEU blend with component ratio 40/60 (3), the same blend treated by water at 70°C (4), and the PHBV/SPEU (60/40) blend also treated by water at 70°C (5). The presence of SPEU in the system leads to an amorphous hallo in the range 22° (see curves 1 and 3), its intensity is increased with the SPEU content.

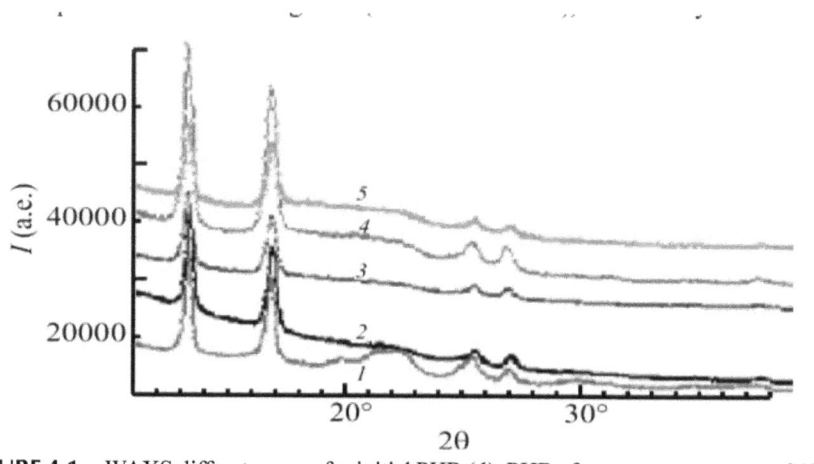

FIGURE 4.1 WAXS diffractograms for initial PHB (*1*), PHB after water treatment at 318 K (*2*), initial blend PHB–SPEU 40:60 (*3*), blend PHB–SPEU 40:60 after water treatment at 343 K (*4*), and PHB–SPEU blend 60:40 after water treatment at 343 K.

The impact of the water medium causes noticeable changes in the crystal structure, namely an increase in the crystallinity degree, the crystallite size growth ,and improving their crystalline structure that manifested as a decrease in the basic interplanar spacings $d020$ and $d110$ (Table 4.1). The ratio of the integrated intensity of the crystalline reflections to the total intensity is 66 percent for the initial samples and 84 percent for samples treated with water. Crystallite sizes, the corresponding lattice parameters, and crystallinity of PHBV are presented in Table 4.1 to summarize the results of X-ray analysis. The above influence of water is similar for hydrophobic PELD with different molecular weights that are subjected to ionizing radiation [13].

TABLE 4.1 Crystalline and thermophysical characterization of PHBV and its Blend

PHB content (%)	Crystallinity (%) WAXS-DSC	Interplanar Spacings (A)	Sizes of Crystallites
100 (initial)	66–58	6,640	215/160
100 (water exposition)	84–59	6,606	255/190
60 (water exposition)	68–58	6,649	183/157
40 (initial)	55–58	6,600	174/149
40 (water exposition)	53–50	6,600	173/145

The exposure of the samples in aquatic environment affects the original PHBV and its composition with SPEU in different ways. The parent biocopolymer shows stability and even improving of the crystalline phase. According to the DSC data, its crystallinity degree is kept constant after the contact with water and approximately is equal to 60 percent. After the water treatment for 4 h and elevated heating (70°C) simultaneously, melting temperature (T_m) shifts to higher temperatures from 175 to 177°C for the native sample and from 172 to 174°C for the sample subjected to the first melting–cooling cycle. These T-scale shifts as well as the analysis by WAXS indicated more perfect organization of PHBV crystals that occurred due to water impact.

For polymer compositions PHBV/SPEU exposed to water at 70°C, the DSC curves of melting show a moderate decrease in crystallinity from 59 to 50 percent. The crystallinity reduction is accompanied by a decrease of T_m from 174°C for the initial sample to 170°C for the PHBV/SPEU blend with 60/40 percent ratio and then to 168°C for the blend with 40/60 percent content. Therefore, the SPEU molecules prevent the crystallization completion of PHBV and, hence, reduce both the quality of structural organization and the crystallinity degree.

4.3   DYNAMICS OF ESR PROBE MOBILITY AND H–D EXCHANGE KINETICS IN D₂O

Recently, we have studied the behavior of spin probes (TEMPO and TEMPOL) in high crystalline PHB at room and elevated temperatures [14]. Water temperature effects on molecular mobility of the ESR probe (TEMPO) in PHBV were studied at two different temperatures: 40°C, close to physiological temperature and 70°C adopted in a number of studies as the standard temperature for accelerated testing of hydrolytic stability of biopolymers [15, 16]. Figure 4.2 shows that the heating of polymer compositions in water for 4 h at ~40°C does not significantly reduce the molecular mobility of the probe in the mixture, as well as does not change the PHBV crystallinity that we have shown above by DSC technique. For samples of PHBV and its blends, the impact of the aquatic environment under more rigorous conditions—that is, at 70°C—is accompanied by the probe mobility increase in comparison with the mobility in the initial polymer systems. These results obtained from ESR spectra reveal plasticizing effect of water molecules, so that segmental mobility of macromolecules in the intercrystalline space is increased, and hence the relaxation processes proceed faster, which leads to an increase in the rotation velocity of the probe TEMPO.

The supplemental feature of SPEU behavior in the blends is an FTIR band intensity decrease for the–NH–fragments of urethane groups which can be represented as a series of kinetic curves. The curves reflect the H–D exchange rate in the PHB/SPEU blends immersed into heavy water. Only those–NH–groups can exchange proton for deuterium that are accessible to the attack by molecules of D2O (see Ref. [17]).

These fragments belong to the amorphous regions and are not included in the domain structures of SPEU. For the blend films, the dependence of exchange rate on the PHBV/SPEU composition has an extreme character (Figure 4.3), and the position of the minimum is in the same concentration range as the minimum value of exchange degree (not shown).

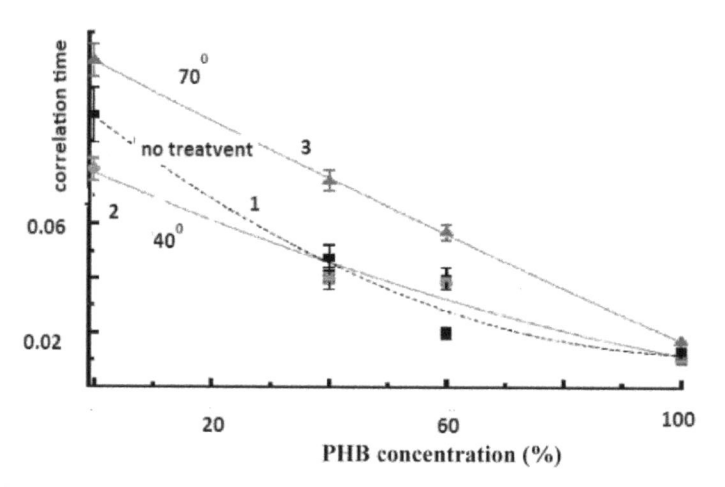

FIGURE 4.2   ESP correlation frequency of TEMPO in the parent polymers (PHB, SPEU) and their blends (1, initial polymer systems; 2, water treatment at 40°C; and 3, water treatment at 70°C).

It was at this ratio of PHBV/SPEU (40:60) that the most ordered structure in the blend was formed and stabilized by intermolecular hydrogen bonds. Figure 4.3 also shows that in this concentration range, there is a maximum oscillation frequency shift ascribed to–NH–fragments in relation to the frequency of–NH–groups in the native SPEU. Such a change in the dynamics of fluctuations is probably due to the formation of the previously mentioned hydrogen bonds in the system.

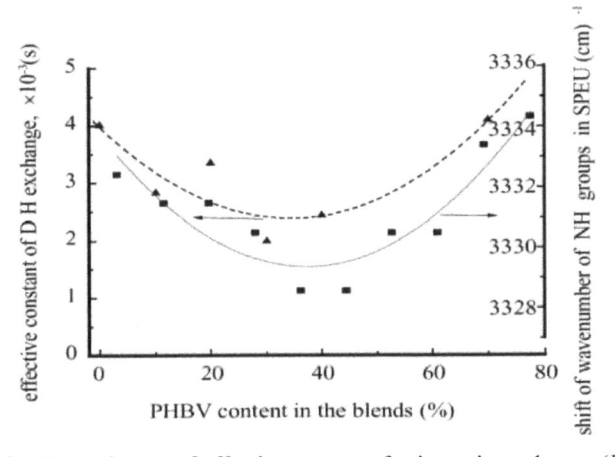

FIGURE 4.3   Dependences of effective constant for isotopic exchange (*left* y-axis) and the shift of wavenumber for –NH–fragment in urethane groups (right y-axis) on PHBV concentration in PHBVSPEU blends.

4.4   CONCLUSIONS

The complex dynamic and structural properties reveal the influence of SPEU on the molecular dynamics and structure of PHBV, by forming intermolecular hydrogen bonds and setting up of energy and steric barriers to PHBV crystallization. In terms of design of new biomedical composites, results presented have a scientific and practical interest to describe their behavior at water temperature exposure.

KEYWORDS

- **Biodegradable composition**
- **Crystallinity**
- **DSC**
- **EPR**
- **X-ray analysis**

REFERENCES

1. Tian, H.; Tang, Z. et al.; *Progress in Polymer Science*. 2012, 37, 237. doi:10.1016/j. progpolymsci.2011.06.004.
2. Suyatma, N. E.; Copinet, A. et al.; *J Polym Environ*. 2004, 12(*1*), 1. doi: 1566-2543/04/0100-01/0.
3. Bonartsev, A. P.; Boskhomodgiev, A. P. et al.; *Mol. Cryst. Liquid Cryst.* 2012, 556 *(1)*, 288.
4. Bagdi, K.; Molnar, K. et al.; *EXPRESS Polym. Lett.* 2011, 5*(5)*, 417. doi: 10.3144/expresspolymlett.2011.41.
5. Shi, R.; Chen, D. et al.; *Int. J. Mol. Sci.* 2009, 10, 4223. doi:10.3390/ijms10104223.
6. Sanche z-Garcia, M. D.; Gimene z, E. et al.; *Carbohydr. Polym.* 2008, 71*(2)*, 235. doi:10.1016/j. carbpol. 2007.05.041.
7. Yang, K-K.; Wang, X.-L. et al.; *J. Ind. Eng. Chem.* 2007, 13*(4)*, 485.
8. Iordanski, A. L.; Rogovina, S. Z. et al.; *Dokl. Phys. Chem.* 2010, 431, Part 2, 60. doi:10.1134/S0012501610040020.
9. Smirnov, A. I.; Belford, R. L. et al.; (Ed. Berliner, L. J.). New York: Plenum Press; 1998, Ch. 3, 83–108.
10. Karpova, S. G.; Iordanskii, A. L. et al.; *Russ. J. Phys. Chem. B.* 2012, 6*(1)*, 72.
11. Bloembergen, S.; Holden, D. A. et al.; Marchessault: Macromolecules. 1989, *22*, 1656.
12. Di Lorenzo, M. L.; Raimo, M. et al.; *J. Macromol. Sci.-Phys.* 2011, B40*(5)*, 639.
13. Selikhova, V. I.; Shcherbina, M. A. et al.; *Polym. Chem.* 2002, 22*(4)*, 605 (in Russian).
14. Kamaev, P. P.; Aliev, I. I. et al.; *Polymer,* 2001, 42, 515.
15. Freier, T.; Kunze, C. et al.; *Biomaterials*, 2002, 23*(13)*, 2649.
16. Artsis, M. I.; Bonartsev, A. P. et al.; *Mol. Cryst. Liq. Cryst.* 2012, 555, 232. doi: 10.1080/15421406.2012.635549.
17. Zaikov, G. E.; Iordanskii, A. L. et al.; Diffusion of Electrolytes in Polymers. Ser. New Concepts in Polymer Science. Utrecht-Tokyo: VSP BV; 1988, 229–231.

CHAPTER 5

NEW BIOLOGICALLY ACTIVE COMPOSITE MATERIALS ON THE BASIS OF DIALDEHYDE CELLULOSE

AZAMAT A. KHASHIROV, AZAMAT A. ZHANSITOV,
GENADIY E. ZAIKOV, and SVETLANA YU. KHASHIROVA

CONTENTS

5.1 INTRODUCTION

Formation and research of systems "polymeric carrier" biologically active substance have recently gained great importance. Such systems find an application as immobilized biocatalysts, bioregulators, and an active form of medicinal substances of the prolonged action.

    In this work, for the first time modification peculiarities of microcrystalline cellulose (MCC) and its oxidized form (dialdehyde cellulose or DAC) guanidine-containing monomers and polymers of vinyl and diallyl series have been studied [1–3]. The structure and some characteristics of used guanidine-containing modifiers are shown in Table 5.1.

TABLE 5.1   Structure and some characteristics of guanidine-containing modifiers of cellulose and dialdehyde cellulose

Modifier	Molecular Weight	Melting Point, °C	Structure
Acrylate guanidine (AG)	131.134	175–176	R=H
Methacrylate guanidine (MAG)	145.160	161–163	R=CH3
N,N-diallylguanidine acetate (DAGA)	199.253	211–212	
N, N-diallylguanidine tri-fluoroacetate (DAGTFA)	253.224	157–158	

5.2   RESULTS AND DISCUSSION

Composite materials were received by treating the microcrystalline cellulose (MCC) or dialdehyde cellulose (DAC), water-soluble monomeric guanidine derivatives (Table 5.1), with subsequent polymerization. Quantity of the monomer/polymer zwitterionic quaternary ammonium cations acrylate and diallyl guanidine derivatives included in MCC or DAC are determined by nitrogen content using elemental analysis.

The results of IR spectroscopic studies show the structural differences of cellulose samples and its modified forms (Figure 5.1).

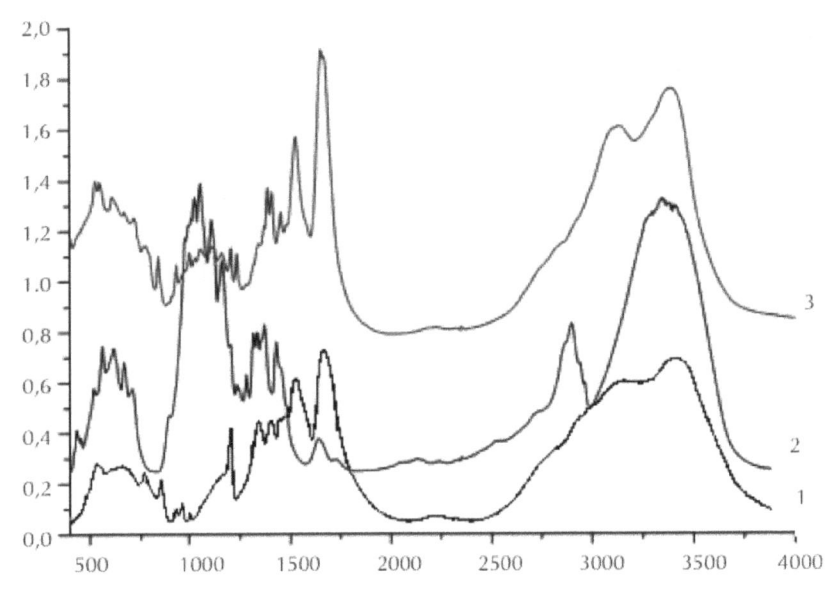

FIGURE 5.1   Comparison of the spectra PMAG (1), DAC–MAG *in Situ* (2), and DAC (3).

For example, the polymerization in DACMAG *in situ* (Figure 5.1, curve 3) in the spectra varies the ratio of the intensity of stripes both cellulose (within 1000–1100 cm^{-1}) and MAG, moreover, the stripe disappears within 860 cm^{-1}, indicating the presence of a double tie. Splitting of the tie C = O ties PMAG within 1250 cm^{-1} takes place, which clearly shows strong mutual influence of the DAC and MAG/PMAG and formation of bimatrix systems. Increasing the intensity of the peak 1660 cm^{-1} in the spectrum of the DAC–MAG formation aldimine connection. Increasing the width of the characteristic absorption stripes in DAC—MAG within 1450–1680 cm⁻¹, probably due to the formation of relatively strong ties with the active MAG–DAC centers.

Comparison of the IR spectra of the composites and DAZ DAZ–DAG and DAZ–DAGTFA demonstrates differences in the spectra of these compounds and indicates the formation of new structures (Figure 5.2).

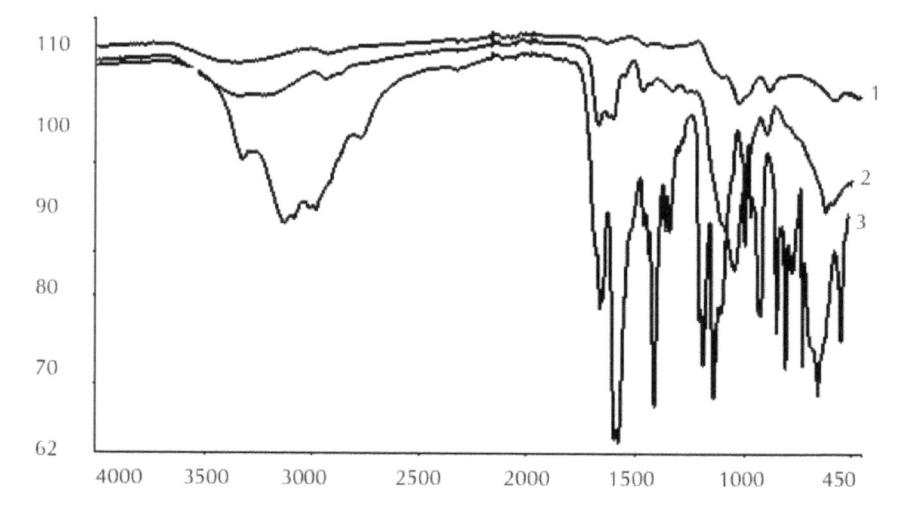

FIGURE 5.2 IR-Spectra DAC (1), DAC–DAGTFA (2), and DAGTFA (3).

In the spectra of the modified DAC, v_{OH} observed increase in absorption from the high frequency, especially in the spectrum of DAC modified DAG. This is due to increasing of the hydroxyls involved in weak hydrogen bonds. Stretching vibrations of C–H bonds of methane and methylene groups DAC appear in 3000–2800 cm⁻¹. In the spectra of the modified DAG and DAGTFA dialdehyde cellulose, these valence oscillations are superimposed on the absorption of the CH_2 groups that are part of diallyl compounds.

This leads to an increase in the intensity of the absorption bands with a frequency of ~ 2900 cm⁻¹. In the area of ~1650 cm⁻¹ peaks appear adsorbed water. Raising polar amino groups comprising the modifier that increases the polarity of the substrate. Increasing the intensity of the peak 1655 cm⁻¹ in the spectrum of DAC–DAG and DAC–DAGTFA may indicate formation aldimine communication giving a signal in this area. In the polymerization in the DAG and DAGTFA, DAC *in situ* peak at 1140 cm⁻¹ present in the DAC disappears. Obviously, the terminal CHO groups DAC and guanidine containing diallyl modifier reacted with each other.

Thus, immobilization AG, MAG, DAG and DAGTFA, DAC between components in the formation of various types of bonds: due to van der Waals' forces, intra and intermolecular coordination and hydrogen bonds, C–C bonds formed during the free radical polymerization *in situ* immobilized AG, MAG, DAG, and DAGTFA, bonds formed during the graft copolymerization of monomer radical salts with DAC and labile covalent aldimine C = N bonds formed by reacting aldehyde groups with amino DAC guanidine-containing compounds.

Composite materials obtained by polymerization and DAG, DAGTFA *in situ* in the inter and intrafibrillar DAC pores dissolve well in water. It can be assumed that the action of pulp and DAGTFA, DAG, a major role in breaking the inter and intramolecular hydrogen bonds play anions CH_3COO^- and SF_3COO^- that form a dialdehyde cellulose stable complexes through its hydroxyl groups. Simultaneously, esterification may occur partly sterically more accessible alcohol groups dialdehyde cellulose. SEM method shows that the dissolution of cellulose sphere-like complex is formed between the components of the solution in which the cellulose macromolecules have coil conformation (Figure 5.3).

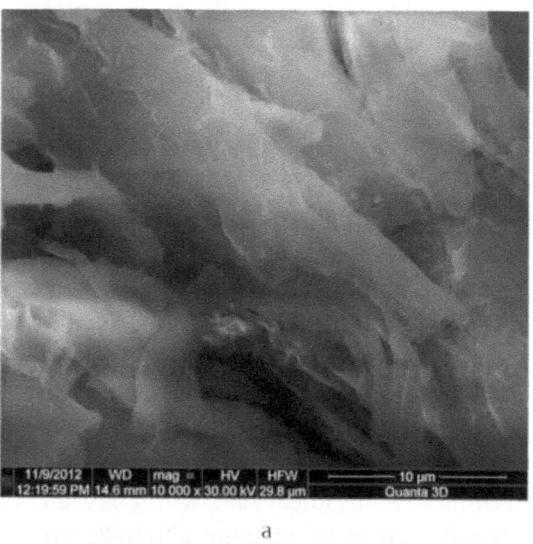

a

b

FIGURE 5.3   SEM images dialdehyde cellulose (a) and complex of dialdehyde cellulose with N,N-diallylguanidine trifluoroacetate (b).

In the process of dissolution of dialdehyde cellulose and DAG, DAGT-FA also acts as acceptors of hydrogen bonds and associated solvent molecules, thereby preventing re-association of cellulose macromolecules.

The biological activity of the synthesized composite materials was investigated, and it was shown that the composite synthesized materials are quite active and have a biocidal effect against Gram-positive (*Staphylococcus aureus* and Gram (*Escherichia coli*) microorganisms. Being the most expressed biocidal activity is shown in the composites with diallyl derivatives of guanidine.

5.3 CONCLUSION

Studies of the structure of the composites by SEM showed that DAG and DAGTFA localized in the surface layers of the composite, which increases the availability of biocidal centers and explains a higher relative activity of these composites.

In the case of dialdehyde cellulose, modified acrylate derivatives, guanidine antibacterial active groups are in the deeper layers of the interfibrillar dialdehyde cellulose, which reduces their bioavailability; and therefore, the DAC–PAG and DAC–PMAG and start to show the bactericidal activity of only 48 h. Slowing down the release rate of the bactericidal agent opens prospect of long-acting drugs with a controlled release bactericide.

KEYWORDS

- Acrylate guanidine
- Cellulose
- IR spectroscopy
- Methacrylate guanidine
- N, N-diallylguanidine acetate
- N, N-diallylguanidine trifluoroacetate
- SEM

REFERENCES

1. Khashirova, S. Yu.; Zaikov, G. E.; Malkanduev, U. A.; Sivov, N. A.; Martinenko, A. I.; Synthesis of new monomers on diallyl gyanidine basis and their ability to radikal (co)

polymerization, J. Biochemistry and Chemistry: Research and developments. New York: Nova Science Publishers Inc; 2003, 39–48.

2. Sivov, N. A.; Martynenko, A. I. Kabanova, E. Yu.; Popova, N. I.; Khashirova, S. Yu.; Ésmurziev, A. M.; Methacrylate and acrylate guanidines: synthesis and properties. *Petrol. Chem.* 2004, *44(1)*, 43–48.

3. Sivov, N. A.; Khashirova, S. Yu.; Martynenko, A. I.; Kabanova, E. Yu.; Popova, N. I.; NMR[1]H spectral characteristics of diallyl monomers derivatives, handbook of condensed phase chemistry. Nova Science Publishers; 2011, 293–301.

CHAPTER 6

MICROHETEROGENEOUS TITANIUM ZIEGLER–NATTA CATALYSTS: THE INFLUENCE OF PARTICLE SIZE ON THE ISOPRENE POLYMERIZATION

ELENA M. ZAKHAROVA, VADIM Z. MINGALEEV,
and VADIM P. ZAKHAROV

CONTENTS

6.1   INTRODUCTION

The formation of highly stereoregular polymers under the action of microheterogeneous Ziegler–Natta catalysts is accompanied by broadening of the polymer MWD [1, 2]. This phenomenon is related to the kinetic heterogeneity of active sites (ACs) [1, 3, 4]. The possible existence of several kinetically nonequivalent ACs of polymerization correlates with the nonuniform particle-size distribution of a catalyst [4]. At present time, much attention is given to study the influence of microheterogeneous catalysts particle size on the properties of polymers [5, 6]. However, almost no detailed study has been made of the effect of particle-size catalysts on their kinetic heterogeneity.

The microheterogeneous catalytic system based on $TiCl_4$ and Al(iso-$C_4H_9)_3$ that is widely used for the production of the *cis*-1,4-isoprene. Our study has shown [7] that the targeted change of the solid phase particle size during the use of a tubular turbulent reactor at the stage of catalyst exposure for many hours is an effective method of controlling the polymerization process and some polymer characteristics of isoprene. It is supposed that the key factor is the interrelation between the reactivity of isoprene polymerization site and the size of catalyst particles on which they localize.

The aim of this study is to investigate the interrelation between the particle size of a titanium catalyst and its kinetic heterogeneity in the polymerization of isoprene.

6.2   EXPERIMENTAL

Titanium catalytic systems (Table 6.1) were prepared through two methods.

Method 1: At 0 or –10°C in a sealed reactor 30–50 mL in volume with a calculated content of toluene, calculated amounts of $TiCl_4$ and Al(iso-$C_4H_9)_3$ toluene solutions (cooled to the same temperature) were mixed. The molar ratio of the components of the catalyst corresponded to its maximum activity in isoprene polymerization. The resulting catalyst was kept at a given temperature (Table 6.1) for 30 min under constant stirring.

TABLE 6.1 Titanium catalytic systems and their fractions used for isoprene polymerization

Catalyst	Labels	Molar ratio of catalyst components			T,°C	Method	Range of particle diameters in fractions of titanium catalysts, μm		
		Al/Ti	DPO/Ti	PP/Al			Fraction I	Fraction II	Fraction III
$TiCl_4$–Al(i-C_4H_9)$_3$	C-1	1	–	-	0	1	0.7–4.5	0.15–0.65	0.03–0.12
						2	–	0.20–0.7	0.03–0.18
$TiCl_4$–Al(i-C_4H_9)$_3$ –DPO	C-2	1	0.15	–	0	1	0.7–4.5	0.15–0.65	0.03–0.12
						2	–	0.15–0.68	0.03–0.12
$TiCl_4$–Al(i-C_4H_9)$_3$ –DPO-PP	C-3	1	0.15	0.15	0	1	–	0.12–0.85	0.03–0.10
					0	2	–	0.15–0.80	0.03–0.12
$TiCl_4$–Al(i-C_4H_9)$_3$ –DPO-PP	C-4	1	0.15	0.15	-10	1	–	0.12–0.45	0.03–0.10
						2	–	0.12–0.18	0.04–0.11
Averaged ranges, μm							0.7–4.5	0.15–0.69	0.03–0.14

Note: DPO—diphenyloxide, PP—piperylene

Method 2: After preparation and exposure of titanium catalysts via Method 1, the system was subjected to a hydrodynamic action via single circulation with solvent through a six-section tubular turbulent unit of the diffuser–confuser design [8] for 2–3 s.

The catalyst was fractionated through sedimentation in a gravitational field. For this purpose, calculated volumes of catalysts prepared through methods 1 and 2 were placed into a sealed cylindrical vessel filled with toluene. In the course of sedimentation, the samples were taken from the suspension column at different heights—a procedure that allowed the separation of fractions varying in particle size.

The titanium concentrations in the catalyst fractions were determined via FEK colorimeter with a blue light filter in a cell with a 50 mm thick absorbing layer. A K_2TiF_6 solution containing 1×10^{-4} g Ti/mL was used as a standard. The catalyst particle-size distribution was measured via the method of laser diffraction on a Shimadzu Sald-7101 instrument.

Before polymerization, isoprene was distilled under the flow of argon in the presence of $Al(iso\text{-}C_4H_9)_3$ and then distilled over a $TiCl_4$–$Al(iso\text{-}C_4H_9)_3$ catalytic system, which provided a monomer conversion of 5–7 per cent. The polymerization on fractions of the titanium catalyst was conducted in toluene at 25°C under constant stirring. In this case, calculated amounts of solvent, monomer, and catalyst were successively placed into a sealed ampoule 10–12 mL in volume. The monomer and catalyst concentrations were 1.5 and 5×10^{-3} mol/L, respectively. The polymerization was terminated via the addition of methanol with 1 per cent ionol and 1 per cent HCl to the reaction mixture. The polymer was repeatedly washed with pure methanol and dried to a constant weight. The yield was estimated gravimetrically.

The MWD of polyisoprene was analyzed via GPC on a Waters GPC-2000 chromatograph equipped with three columns filled with a Waters microgel (a pore size of 103–106 A) at 80°C with toluene as an eluent. The columns were preliminarily calibrated relative to Waters PS standards with a narrow MWD (M_w/M_n = 1.01). The analyses were conducted on a chromatograph, which allows calculations with allowance for chromatogram blurring. Hence, the need for additional correction of chromatograms was eliminated.

The microstructure of polyisoprene was determined via high-resolution ^{1}H NMR spectroscopy on a Bruker AM-300 spectrometer (300 MHz). The MWD of *cis*-1,4-polyisoprene obtained under the aforementioned experimental conditions, $q_w(M)$, were considered through the following equation:

$$q_w(M) = \int_0^\infty \Psi(\beta) M \beta^2 \exp(-M\beta) d\beta$$

where β is the probability of chain termination and $\psi(\beta)$ is the distribution of active site over kinetic heterogeneity, M is current molecular weight.

As shown earlier [9], Eq. (6.1) is reduced to the Fredholm integral equation of the first kind, which yields function $\psi(\beta)$ after solution via the Tikhonov regularization method. This inverse problem was solved on the basis of an algorithm from Ref. [9]. As a result, the function of the distribution over kinetic heterogeneity in $\psi(\ln\beta)$—$\ln M$ coordinates with each maximum related to the functioning of AC of one type was obtained.

6.3 RESULTS

After mixing of the components of the titanium catalyst, depending on its formation conditions, particles 4.5 μm to 30 nm in diameter, which are separated into three arbitrary fractions, are formed (Table 6.1).

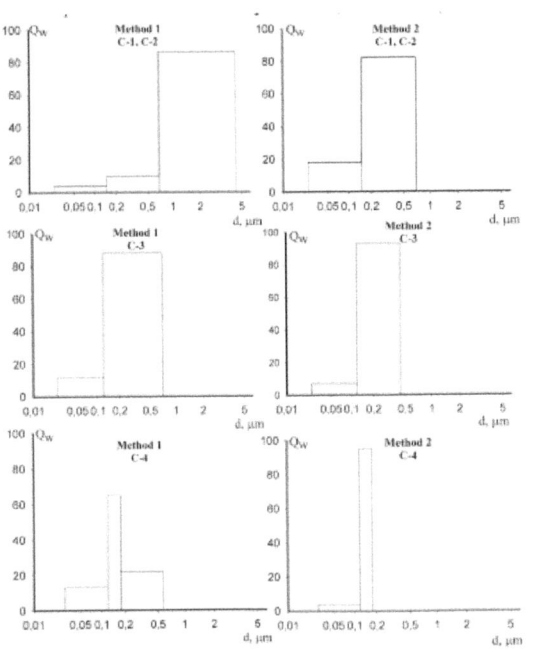

FIGURE 6.1 Fractional compositions of titanium catalyst C-1−C-2 (Table 6.1).

During the formation of catalyst C-1 via Method 1, the fraction composed of relatively coarse particles, fraction I, constitutes up to 85 per cent (Figure 6.1). In Method 2, the hydrodynamic action on the titanium catalyst formed under similar conditions results in an increase in the content of fraction II. Analogous trends are typical of catalyst C-2. The catalyst modification with piperylene additives, catalyst C-3, is accompanied by the disappearance of fraction I and an increase in the content of fraction II (Figure 6.1), as was found during the hydrodynamic action on C-1. The hydrodynamic action on a two-component catalyst is equivalent to the addition of piperylene to the catalytic system. The preparation of catalytic complex C-3 via Method 2 results in narrowing of the particle-size distribution of fraction II owing to disintegration of particles 0.50–0.85 μm in diameter (Figure 6.1). The reduction of the catalyst exposure temperature to −10°C (catalyst C-4) is accompanied by further disintegration of fraction II (Figure 6.1). In this case, the content of particles 0.19–0.50 μm in diameter decreases to 22 per cent with predominance of particles 0.15–0.18 μm in diameter. The formation of C-4 via Method 2 results in additional dispersion of particles of fraction II, with the content of particles 0.15–0.18 μm in diameter attaining 95 per cent.

The content of the finest catalyst particles in the range 0.03–0.14 μm, fraction III, attains 5–12 per cent and is practically independent of the catalyst formation conditions. The most considerable changes are shown by particles 0.18–4.50 μm in diameter. Particles of fraction I are easily dispersed as a result of the hydrodynamic action in turbulent flows and the addition of catalytic amounts of piperylene, and their diameter becomes equal to that of particles from fraction II. The decrease in the catalyst exposition temperature from 0 to −10°C with subsequent hydrodynamic action leads to a more significant reduction of particle size and the formation of a narrow fraction.

Isolated catalyst fractions differing in particle size were used for isoprene polymerization. The *cis*-1,4-polymer was obtained for all fractions, regardless of their formation conditions. The contents of *cis*-1, 4 and 3, 4 units were 96–97 and 3–4 per cent, respectively. Coarse particles (fraction I) are most active in isoprene polymerization (method 1) on different fractions of C-1 (Figure 6.2).

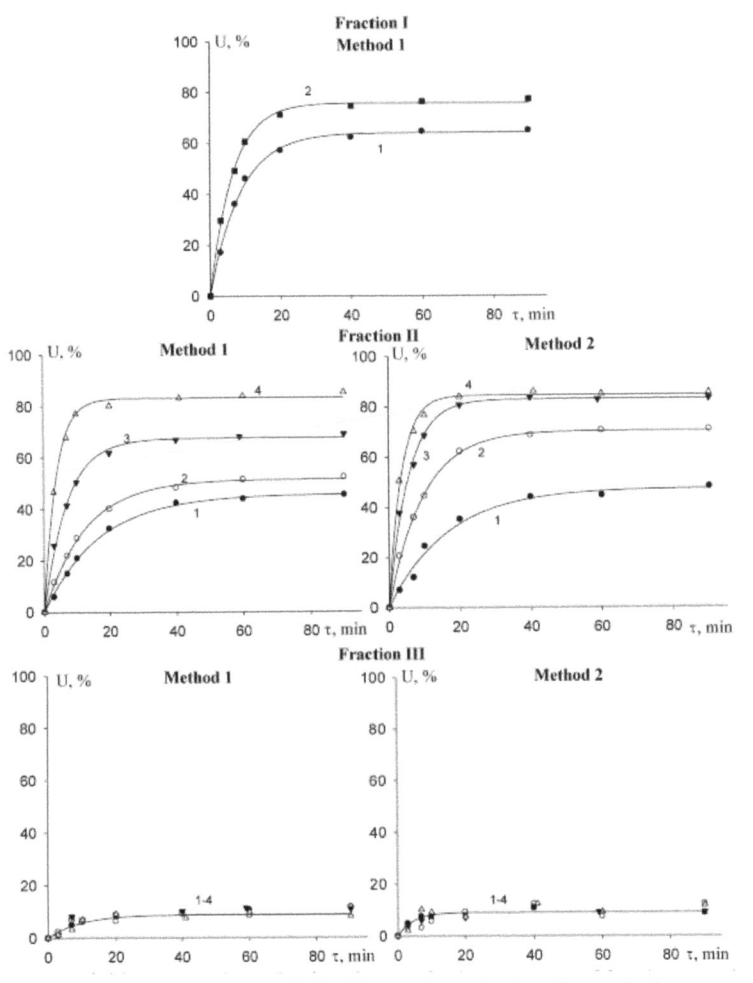

FIGURE 6.2 *cis*-1,4-Polyisoprene yields U vs. polymerization times τ in the presence of fractions particles of titanium catalysts (*1*) C-1, (*2*) C-2, (*3*) C-3, and (*4*) C-4.

As the particle size of C-1 decreases, its activity drops significantly. The catalyst modification with diphenyloxide (C-2) has practically no effect on the fractional composition, but the activities of different catalyst fractions change. The most marked increase in activity was observed for fraction I. Catalyst C-3 prepared via Method 1 comprises two fractions, with fraction II having the maximum activity. The decrease of the catalyst exposition temperature to –10°C (C-4) result in further increase in the rate of isoprene polymerization on particles of fraction II.

The hydrodynamic action increases the content of fraction II in C-1, but its activity in isoprene polymerization does not increase (Figure 6.2). For C-2, the analogous change in the fractional composition is accompanied by an increased activity of fraction II (Method 2). The addition of piperylene to C-3 results in a stronger effect on the activity of fraction II under the hydrodynamic action. The change of the hydrodynamic regime in the reaction zone does not affect the activity of fine particles of catalyst fraction III. Isoprene polymerization in the presence of fraction III always has a low rate and a cis-1,4-polyisoprene yield not exceeding 7–12 per cent.

During polymerization with C-1 composed of fraction I (Method 1), the weight-average molecular mass of polyisoprene increases with the process time (Table 6.2). Polyisoprene prepared with catalyst fraction II has a lower molecular mass. A more considerable decrease in the weight-average molecular mass is observed during isoprene polymerization on the finest particles of fraction III, with M_w being independent of the polymerization time. The width of the MWD for polyisoprene obtained on fraction I increases with polymerization time from 2.5 to 5.1. The polymer prepared in the presence of catalyst fraction III shows a narrow molecular mass distribution ($M_w/M_n \sim 3.5$). The addition of DPO is accompanied by an increase in the weight-average molecular mass of polyisoprene obtained in the presence of fraction I. Under a hydrodynamic action (Method 2) on catalysts C-1 and C-2, polyisoprene with an increased weight-average molecular mass is formed. During the addition of piperylene to the catalyst (C-3), M_w of polyisoprene increases to values characteristic of C-2 formed via method 2. The formation of the catalytic system $TiCl_4$-$Al(iso$-$C_4H_9)_3$-DPO-piperylene via method 1 at $-10°C$ results in substantial increases in the activity of the catalyst and the reactivity of the active centers of fraction II. The polymerization of isoprene on catalyst fraction III, regardless of the conditions of catalytic system formation (electrondonor additives, exposure temperature, hydrodynamic actions), yields a low-molecular-mass polymer with a polydispersity of 3.0–3.5.

TABLE 6.2 Molecular-mass characteristics of *cis*-1,4-polyisoprene. 1 and 2 are the methods of catalyst preparation

C	τ Min	Fraction I				Fraction II				Fraction III			
		$M_w \times 10^{-4}$		M_w/M_n		$M_w \times 10^{-4}$		M_w/M_n		$M_w \times 10^{-4}$		M_w/M_n	
		1	2	1	2	1	2	1	2	1	2	1	2
C-1	3	17.1		2.6		43.2	43.6	4.6	4.8	25.4	21.4	3.7	3.8
	20	56.3		4.2		46.5	57.5	4.0	5.9	23.7	22.6	3.4	3.7
	40	64.9		4.6		47.4	62.7	4.1	5.8	27.6	28.9	3.6	3.6
	90	72.2		5.1		54.3	61.4	4.3	6.2	26.8	22.4	3.7	3.6
C-2	3	24.3		3.3		39.5	54.8	4.3	4.9	21.1	22.3	3.8	3.7
	20	69.7		3.8		41.2	67.5	3.9	4.6	24.6	24.6	3.7	3.4
	40	78.4		3.7		42.6	72.8	4.4	4.3	26.3	22.4	3.3	3.5
	90	84.6		4.0		58.0	73.2	4.3	4.2	24.7	24.3	3.5	3.3
C-3	3					66.2	60.5	3.9	3.8	22.4	20.0	3.8	3.6
	20					75.4	71.1	3.7	3.5	28.1	21.3	3.7	3.7
	40					79.6	78.8	4.1	4.0	21.5	23.6	3.8	3.8
	90					81.6	79.6	3.9	3.8	23.2	26.8	3.6	3.3
C-4	3					81.2	72.7	3.6	3.7	25.4	20.3	3.8	3.6
	20					62.7	51.3	3.3	3.2	24.8	21.6	3.7	3.7
	40					57.6	44.3	3.1	3.2	23.2	26.8	3.6	3.6
	90					61.4	48.2	3.2	3.4	19.6	21.5	3.1	3.3

The solution of the inverse problem of the formation of the MWD in the case of *cis*-1,4-polyisoprene made it possible to obtain curves of the active site distribution over kinetic heterogeneity (Figures 6.3–6.6). As a result of averaging of the positions of all maxima, three types of polymerization site that produce isoprene macromolecules with different molecular masses were found: type A (lnM = 10.7), type B (lnM = 11.6), and type C (lnM = 13.4).

The polymerization of isoprene in the presence of fractions I and II of C-1 (Method 1) occurs on site of types B and C (Figure 6.3). The polymerization in the presence of fraction III proceeds on active site of type A only. The single site and low activity character of the catalyst composed of particles of fraction III is typical of all the studied catalysts and all the methods of their preparation. Thus, these curves are not shown in subsequent figures. Likewise, catalyst C-2 prepared via Method 1 features the presence of type B and type C site in isoprene polymerization on fractions I and II (Figure 6.4). The hydrodynamic action (Method 2) on C-1 accompanied by dispersion of particles of fraction I does not change the types of site of polymerization on particles of fraction II (Figure 6.5). A similar trend is observed during the same action on a DPO-containing titanium catalyst in turbulent flows (Figure 6.5).

In the presence of piperylene, the main distinction of function $\psi(\ln\beta)$ relative to the distributions considered above is a significantly decreased area of the peak due to the active site of type B (Figure 6.6). The decrease of the catalyst exposition temperature to $-10°C$ (C-4, Method 1) allows the complete "elimination" of active site of type B (Figure 6.6). With allowance for the low content of fraction III, it may be concluded that under these conditions, a single-site catalyst is formed. In the case of the hydrodynamic action on C-4, the particles of fraction II contain active centers of type C with some shift of unimodal curve $\psi(\ln\beta)$ to smaller molecular masses (Figure 6.6).

6.4 DISCUSSION

The particles of the titanium catalyst 0.03–0.14 μm in diameter, regardless of the conditions of catalytic system preparation, feature low activity in polyisoprene synthesis, and the resulting polymer has a low molecular mass and a narrow MWD. The molecular mass characteristics of polyisoprene and the activity of the catalyst comprising particles 0.15–4.50 μm in diameter depend to a great extent on its formation conditions.

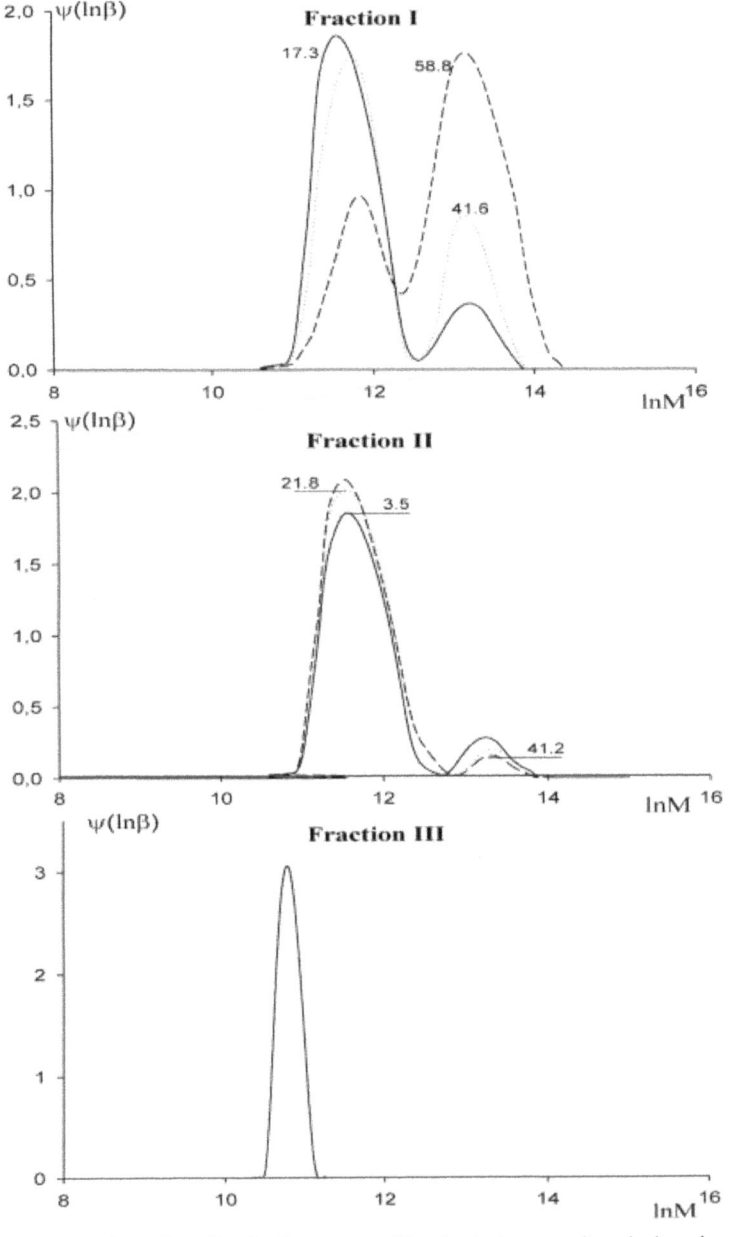

FIGURE 6.3 Active site distributions over kinetic heterogeneity during isoprene polymerization on fractions C-1. Method 1. Here and in figures 6.4–6.5 numbers next to the curves are conversions (%).

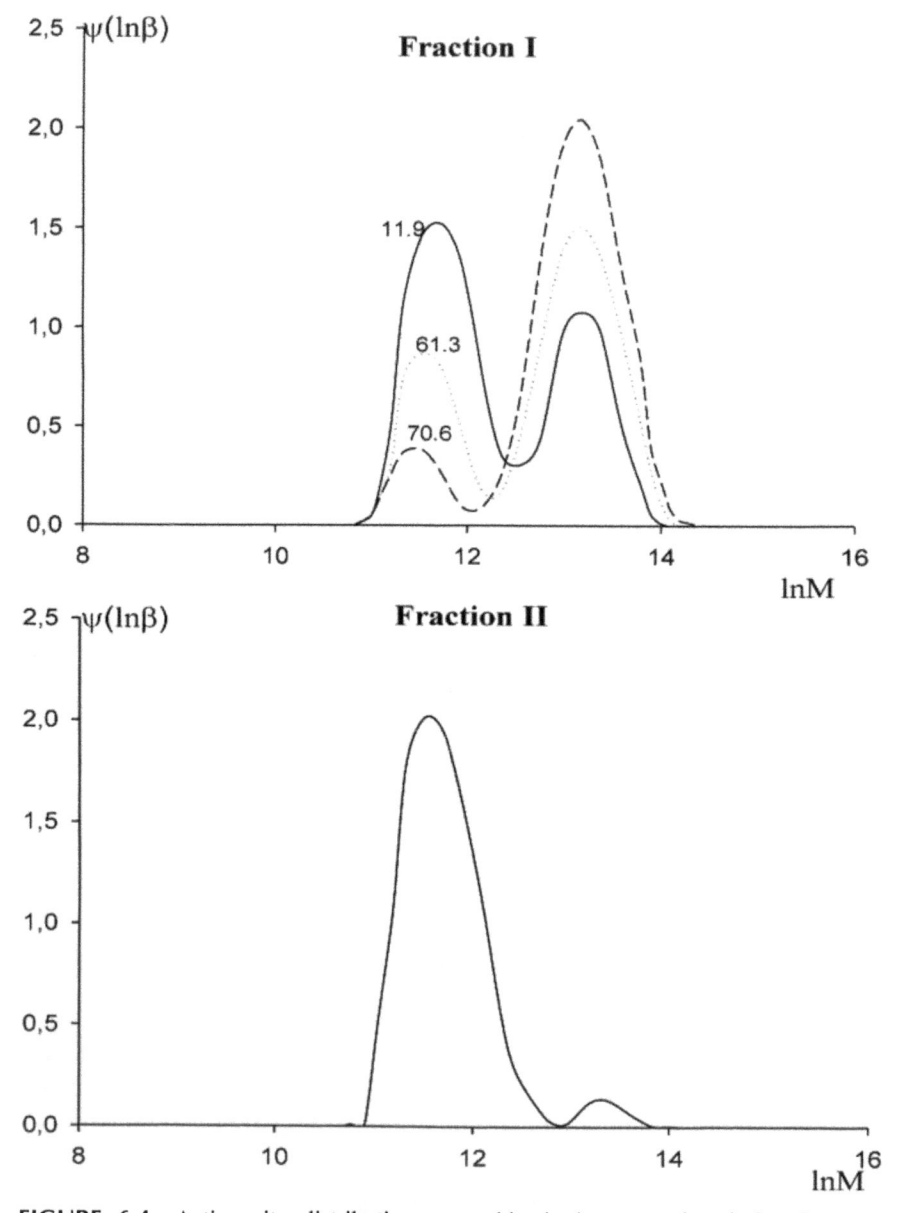

FIGURE 6.4 Active site distributions over kinetic heterogeneity during isoprene polymerization on fractions C-2. Method 1.

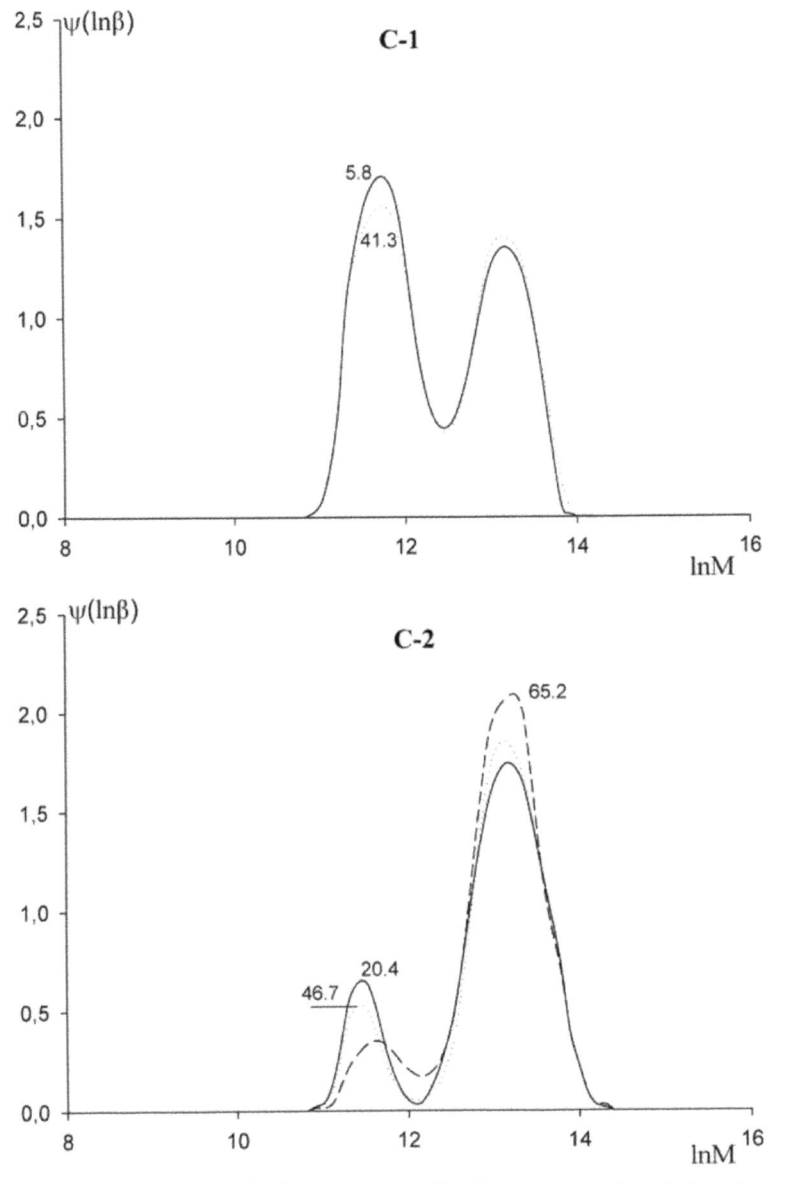

FIGURE 6.5 Active site distributions over kinetic heterogeneity during isoprene polymerization on particles of fraction II of C-1 and C-2. Method 2.

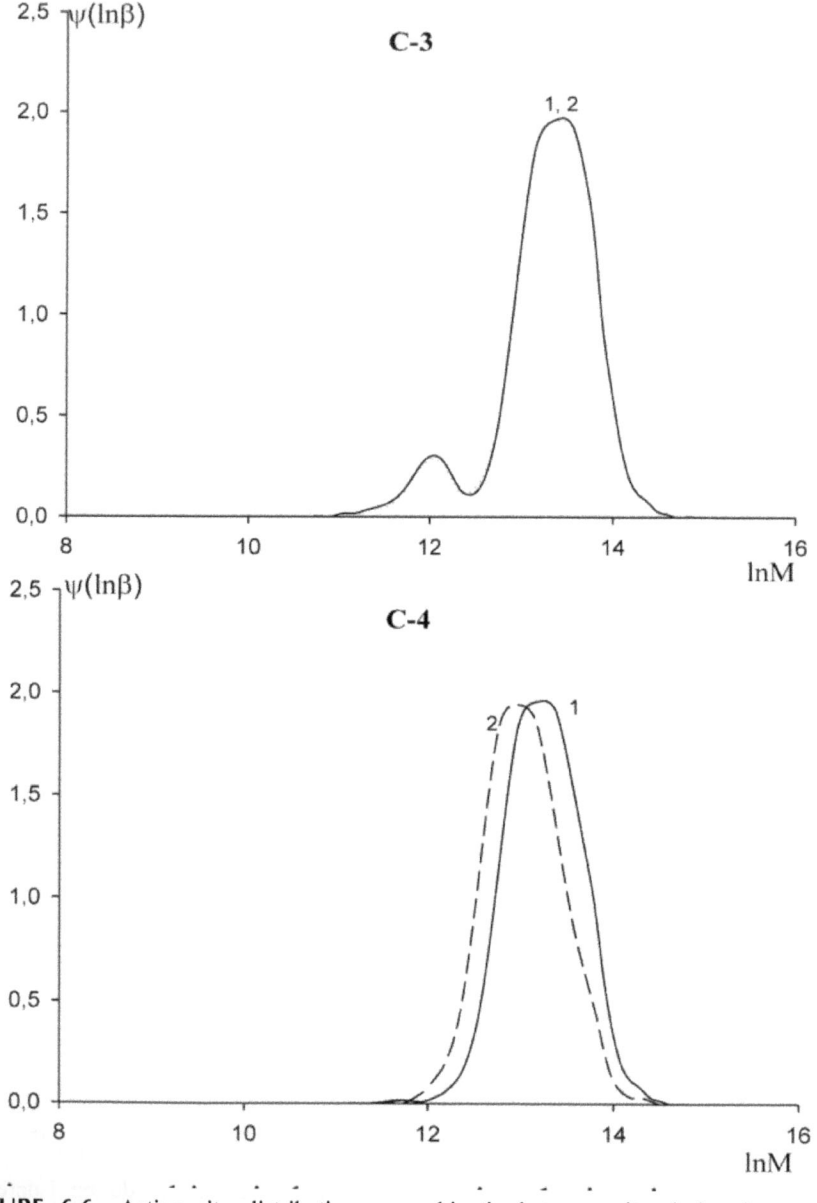

FIGURE 6.6 Active site distributions over kinetic heterogeneity during isoprene polymerization on particles of fraction II of C-3 and C-4. Methods 1-1, Method 2-2.

As shown earlier [10, 11], the region of coherent scattering for particles based on $TiCl_3$ spans 0.003–0.1 µm, a range that corresponds to the linear size of the minimum crystallites. Coarser catalyst particles are aggregates of these minimum crystallites. This circumstance makes it possible to suggest that the fraction of catalyst particles 0.03–0.14 µm in diameter that was isolated in this study is a mixture of primary crystallites of β-$TiCl_3$ that cannot be separated via sedimentation. The fractions of catalyst particles with larger diameters are formed by stable aggregates of 2–1100 primary crystallites. There is sense in the suggestion that the elementary crystallites are combined into larger structures via additional Al–Cl bonds between titanium atoms on the surface of a minimum of two elementary crystallites, that is $(Ti)_1$–Cl–Al–Cl–$(Ti)_2$.

Similar structures can be formed with the participation of AlR_2Cl and $AlRCl_2$, which are present in the liquid phase of the catalyst. Trialkyl aluminum AlR_3 is incapable of this type of bonding. Thus, the structure of the most alkylated Ti atom (in the limit, a monometallic center of polymerization), which has the minimum reactivity, should be assigned to the active centers localized on particles 0.03–0.14 µm in diameter [9]. On particles 0.15–4.50 µm in diameter in clusters of primary crystallites, highactivity bimetallic centers with the minimum number of Ti–C bonds at a Ti atom are present. Thus, the experimental results obtained in this study show that the nature of the polymerization center resulting from successive parallel reactions between the pristine components of the catalytic system determines the size of the titanium catalyst particles and, consequently, their activity in isoprene polymerization.

6.5 CONCLUSION

We first examined the isoprene polymerization on the fractions of the titanium catalyst particles that were isolated by sedimentation of the total mixture. The results obtained allow considering large particle as clusters that are composed of smaller particles. In the process of polymerization or catalyst preparation, the most severe effects are large particles (clusters). This result in the developing process later on substantially smaller as compared with initial-size particles. These particles are fragments of clusters which are located over the active centers of polymerization. Note that the stereo specificity is not dependent on the size of the catalyst particles.

Hypothesis about clusters agrees well with the main conclusions of this paper:

I.   Isolated the fraction of particles of titanium catalyst $TiCl_4$–Al(iso-$C_4H_9)_3$: I—0.7–4.5 µm, II—0.15–0.68 µm, III—0.03–0.13 µm. With decreasing particle size decreases the rate of polymerization the molecular weight and width of the molecular weight distribution. Hydrodynamic impact leads to fragmentation of large particles of diameter greater than 0.5 µm.

II. Isoprene polymerization under action of titanium catalyst is occurs on three types active sites: type A—lnM = 10.7; type B—lnM = 11.6; and type C—lnM = 13.4. Fractions I and II particles contain the active site of type B and C. The fraction III titanium catalyst is represented by only one type of active sites producing low-molecular-weight polymer (lnM = 10.7).

III. The use of hydrodynamic action turbulent flow, doping DPO and piperylene, lowering temperature of preparation of the catalyst allows to form single site catalyst with high reactivity type C (lnM = 13.4), which are located on the particles of a diameter of 0.15–0.18 µm.

ACKNOWLEDGMENTS

This study was financially supported by the Council of the President of the Russian Federation for Young Scientists and Leading Scientific Schools Supporting Grants (project no. MD-4973.2014.8).

KEYWORDS

- Active sites
- Isoprene polymerization
- Particles-size effect
- Single-site catalysts
- Ziegler–Natta catalyst

REFERENCES

1. Kissin, Yu. V.; *J. Catal.* **2012,** *292,* 188–200.
2. Hlatky, G. G.; *Chem. Rev.* **2000,** *100,* 1347–1376.
3. Kamrul Hasan, A. T. M.; Fang, Y.; Liu, B.; Terano, M.; *Polymer,* **2010,** *51,* 3627–3635.
4. Schmeal, W. R.; Street, J. R.; *J. Polym. Sci: Polym. Phys. Ed.* **1972,** 10, 2173–2183.
5. Ruff, M.; Paulik, C.; *Macromol. React. Eng.* **2013,** 7, 71–83.
6. Taniike, T.; Thang, V. Q.; Binh, N. T.; Hiraoka, Y.; Uozumi, T.; Terano, M.; *Macromol. Chem. Phys. 212,* **2011,** 723–729.
7. Morozov, Yu. V.; Nasyrov, I. Sh.; Zakharov, V. P.; Mingaleev, V. Z.; Monakov, Yu. B.; *Russ. J. Appl. Chem.,* **2011,** *84,* 1434–1437.
8. Zakharov, V. P.; Berlin, A. A.; Monakov, Yu. B.; Deberdeev, R. Ya.; Physicochemical fundamentals of rapid liquid phase processes. Moscow: Nauka; **2008,** 348 p.
9. Monakov, Y. B.; Sigaeva, N. N.; Urazbaev, V. N.; Active sites of polymerization. Multiplicity: stereospecific and kinetic heterogeneity. Leiden: Brill Academic, **2005,** 397 p.
10. Grechanovskii, V. A.; Andrianov, L. G.; Agibalova, L. V.; Estrin, A. S.; Poddubnyi, I. Ya.; *Vysokomol. Soedin.,* **1980,** Ser. A *22,* 2112–2120.
11. Guidetti, G.; Zannetti, R.; Ajò, D.; Marigo, A.; Vidali, M.; *Eur. Polym. J.* **1980,** *16,* 1007–1015.

CHAPTER 7

THE ROLE AND MECHANISM OF BONDING AGENTS IN COMPOSITE SOLID PROPELLANTS

S. A. VAZIRI, S. M. MOUSAVI MOTLAGH, and M. HASANZADEH

CONTENTS

7.1   INTRODUCTION

Composite propellants are the basic component for development and production of modern propellants. Comparison of composite propellants with double-base propellants indicated that composite propellants have remarkable properties, such as high performance, high density, and superior physical properties for producing large grain; therefore, they are excellent candidates for application in military systems such as strategic ballistic missiles as well as space shuttle boosters in space systems. Bonding agents, as an extremely important component in solid rocket propellants, may affect the processability, mechanical and ballistic properties, safety, aging, and temperature cycles of propellants. Recently, nitramine used as an oxidizer in propellant due to its potential advantages [1].

Allen [2] investigated the application of water-soluble animal collagen proteins for cyclotrimethylene—trinitramine (RDX) and cyclotetramethylene—tetranitramine (HMX) as a bonding agent. In the study on the effect of tensile strain rate on the mechanical properties of HTPB-based fuels by [3], it was demonstrated that the elongation at break increases with increasing strain rate. However, with increasing solids, tensile strength increases and elongation at break decreases. [4] indicated that [tris (2-hydroxyl) isocyanurate] can be used as bonding agent for CTPB/Al/RDX composite propellants system. They investigated the interaction of several bonding agents based on isocyanurate and found that the oxidizers form hydrogen bonds with [tris (2-hydroxyl) isocyanurate] and [tris (2-carboxy-ethyl) isocyanurate]. The investigation was performed on several isocyanurate bonding agents with AP and several binders by using FTIR analytics. It is found that the oxidizers interact with bonding agent [tris (2-hydroxyl) isocyanurate] and [tris (2-carboxy-ethyl) isocyanurate] through hydrogen bond. Moreover, it is found that RDX interacts with both bonding agents, whereas HTPB have only interacted with [tris (2-carboxy-ethyl) isocyanurate] [4].In another work, [5] studied the effect of RDX on tensile properties and elongation of the AP/HTPB composite propellants. They found that in the presence of RDX, elongation at maximum stress has increased about 90 per cent, at room temperature. RDX also reduces the burning rate and improves the processability of these propellants. The result showed that increasing the temperature from 25°C to 55°C leads to an increase in elongation and tensile strength and slightly decreases the elastic modulus. Moreover, the mechanical properties increase with increasing strain rate from 1 to 1,000 mm/min [5]. [6] studied the relationship between micro-structure and mechanical properties of HTPB-based fuels. It is found that mechanical properties of HTPB-based fuel are closely related to its micro-structure. They explained that the structure of composite solid propellants and also their mechanical properties are influenced by a great number of parameters including binder/solid interface, size, shape and distribution of the solid particles, and the quantity of binder matrix. Composite solid propellants containing nitramine with RDX or HMX offer many advantages such as combustion with low-flame temperature and low molecular weight. Generally, in composite propellants formulation, single-crystalline nitramine propellants such as HMX or RDX mixed with polymeric binder [7].

To improve the mechanical properties of composite solid propellants containing nitramine, the addition of plasticizer, cross-linking materials, suitable catalyst and curing agent and also the addition of the bonding agent have been proposed. According to literature, the bonding agents play a significant role in increasing adhesions between the binder and the solid phase, and therefore improving the mechanical properties of solid propellants [1, 8].

7.2 THE MECHANISM OF BONDING AGENTS IN IMPROVING MECHANICAL PROPERTIES

It is well known that the main components of the composite solid propellants are oxidizer, metal fuel, and polymeric binders (Figure 7.1). Since composite solid propellants are not homogeneous, the oxidizer and metal fuel are suspended in the binder. Therefore, the presence of solid particles (such as AP and Al) in composite solid propellants can lead to grain separation [4, 8].

The addition of bonding agents increased the strength of composite by producing chemical or physical bond between binder and oxidizer [8, 10]. The mechanism of improved mechanical performance of composite solid propellants is based on the absorption of bonding agent on the surface of oxidizer and its chemical reaction with binder (Figure 7.2). The compatibility of bonding agent with solid particles and having suitable functional groups for chemical bond formation are two essential features of bonding agent [8].

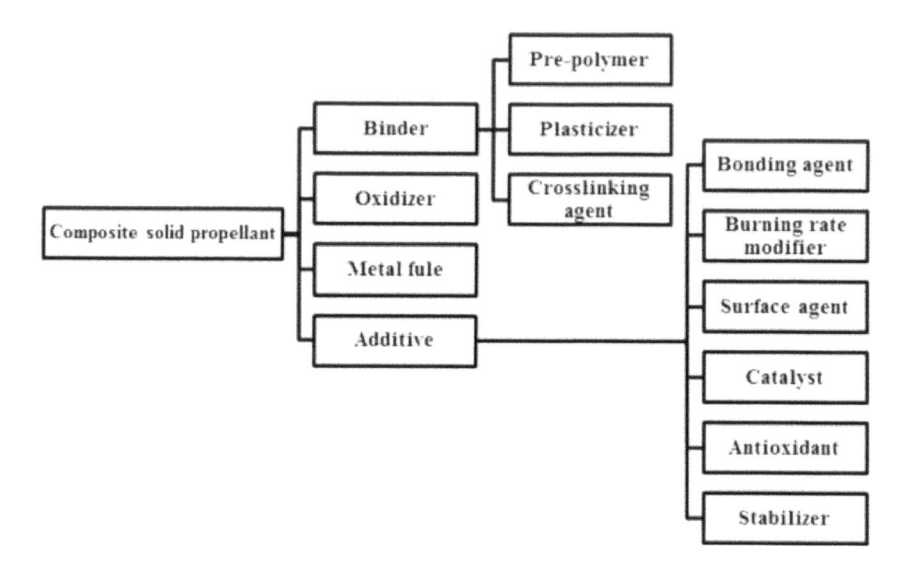

FIGURE 7.1   Classification of composite solid propellant components [1, 9].

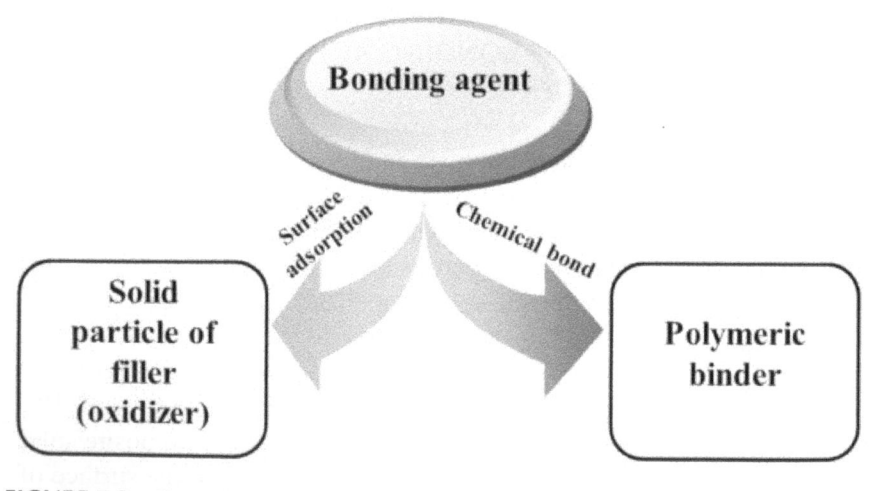

FIGURE 7.2   General mechanism of bonding agents [8].

7.3 APPROPRIATE BONDING AGENTS FOR PROPELLANTS CONTAINING NITRAMINE

There are convenient bonding agents for AP such as polyamines, aziridine, oxazoline and amine salts, schiff base, tetra-alkyl titanate, and vinyl ethers [11–16]. For nitramine fillers (such as RDX), bonding agents include di (2-hydroxyethyl) dimethylhydantoin (DHE), vinyltriethoxysilane (VTEO), triethanolamine (TEA), maleic anhydrate–terminated poly butadiene (MPB) [17–19]. In recent years, hydantoin are known as most popular bonding agent for RDX and HMX in solid propellant composites. Moreover, hydantoin has been used as bonding agent in KP, AN, and AP oxidizers. In this section, the effective bonding agents used for improving mechanical properties of composite solid propellants containing nitramine are discussed in detail [18].

7.3.1 NEUTRAL POLYMERIC BONDING AGENT

Alkanolamines, alkanolamides, polyamines, Dantcol, and amine salts are typical examples of bonding agents for propellants, with relatively nonpolar binders such as polybutadiene glycol. These small polar molecules, due to the nonpolarity of binder matrix, undergo adsorption on polar solid particles. However, these bonding agents are no longer effective in a system where polar binders such as nitro and nitratoplasticizers are used. This is because these bonding agents are too soluble in the polar submix. When the solid particles are nitramine crystals (such as HMX), adsorption becomes even less likely since they have cohesive energy density close to that of the binder matrix. However, in situations where the polarity of the submix approaches that of the solid particles, this criterion is difficult to satisfy [10].

The polar plasticizers not only compete with the bonding agent for adsorption on the solid surfaces but also dissolve the bonding agent extensively. In some cases, these polar plasticizers are even known to partially dissolve the HMX particles. Although the bonding agents with acidic or basic molecules tend to absorb more solid particles, they are usually detrimental to the cure or the aging stability of binders containing nitrato, nitro, or azido groups. For such binder systems, bonding agents should not only be neutral and polymeric, that is, a neutral polymeric bonding agent

(NPBA), but they should also have a high affinity for the solid particles. Moreover, NPBA should contain many hydroxyl groups per molecules, that are enough available to undergo crosslinking with the isocyanate and also form primary bonds with the binder matrix. Polymers containing one or more functionalities such as nitro, nitrato, cyano, sulfone, amide, sulfonamide, substituted amides such as cyanoethylated amides, and ammonium salt groups are some of the most examples of such bonding agents. This bonding agents should have an average molecular weight in the range from 3,000 to 5,00,000 (preferably between 5,000 and 1,00,000), and have ato 100 hydroxyl groups per polymer molecule [10, 20].

It has been proven that the strength and initial modulus of high-energy propellant containing NG, PEG, and HMX can be improved by addition of small amount of NBPA bonding agent during the mixing process. Although deviation from neutrality often causes poor cure and/or impairs stability of the binder components, it is known that the use of basic or acidic bonding agents may improve the degree of adsorption [10].

Addition of a small amount (0.2–0.6 % of the propellant formulation) of neutral polymeric bonding agent leads to filler reinforcement. For example, it is found that HMX/PEG with 0.2 wt. per cent of NPBA will result in a fivefold increase of elongation in comparison to their counterparts without NPBA. Figures (7.3) shows the stress–strain curves of HMX particles coated by NBPA bonding agents [10].

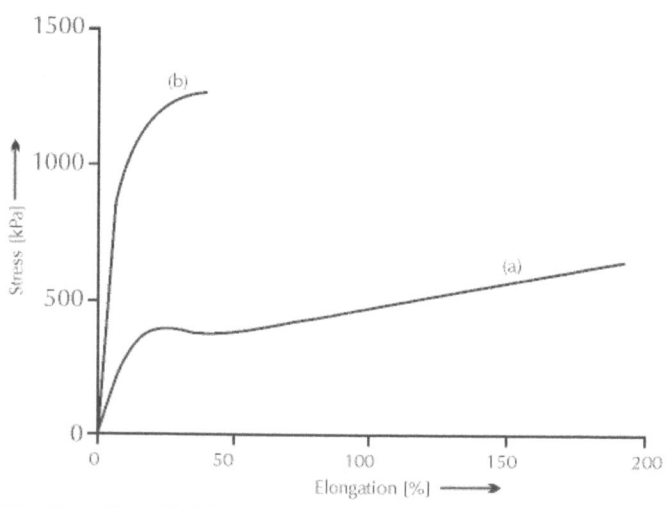

FIGURE 7.3　The effect of NBPA coating on mechanical properties of HMX particles: (a) uncoated HMX and (b) coated HMX [10].

Figure (7.4) shows the mechanical properties of HMX-filled and un-filled composite. It can be noted that incorporation of 0.2 wt. per cent of NPBA gives a material with significantly greater strength than the unfilled binder [10].

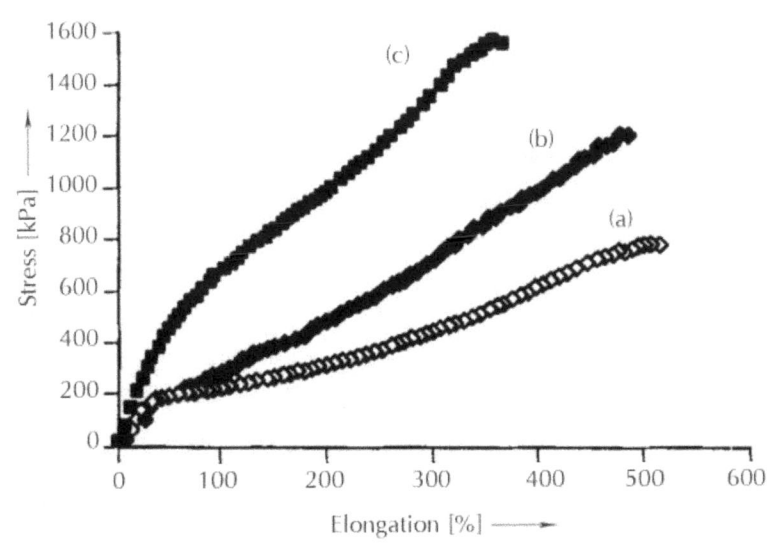

FIGURE 7.4   Stress-strain curve of HMX-filled and unfilled composite: (a) 40 per cent HMX without NPBA, (b) without HMX and NPBA, and (c) 40 per cent HMX and 0.2 per cent NPBA [10].

Figure (7.5) shows another demonstration of the effectiveness of NPBA in composite with higher plasticizer and solid contents, and also larger particle size in comparison with that of illustrated in Figure (7.4). It is found that the incorporation of 0.6 wt. per cent of NPBA into the PEG/TMETN/NP composite with 50 per cent HMX, would result the stress–strain curve approaches the curve of the composite containing coated HMX particles [10].

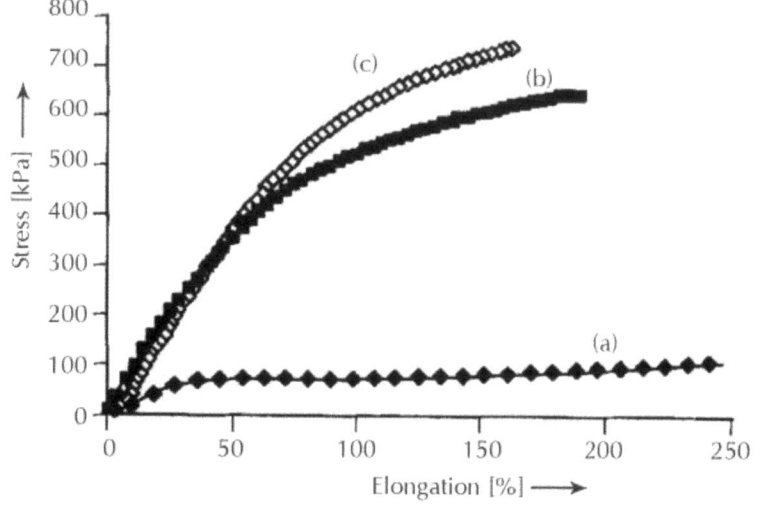

FIGURE 7.5   Effect of NPBA on mechanical properties of composite, (a) without HMX and NPBA, (b) 50 per cent HMX, 0.6 per cent NPBA, and (c) 50 per cent HMX without NPBA [10].

In another study, [11] investigated the mechanisms of NPBA in a HMX/NG/PEG composite. It is found that during the curing process, bonding agent tied with a PEG binder, and thereby the initial modulus will increase. The chemical reaction of NBPA and PEG binder are shown in Figure (7.6) [10, 11].

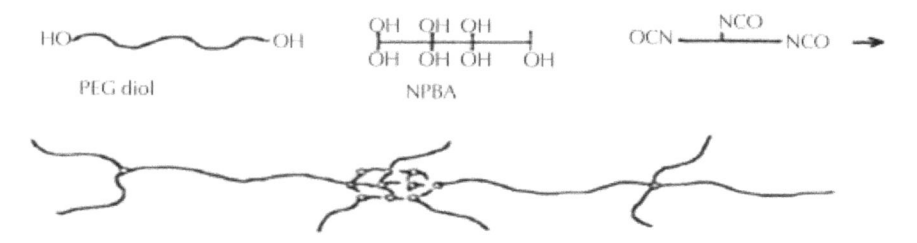

FIGURE 7.6   The chemical reaction of NBPA bonding agent and PEG binder [11].

7.3.2 SILANE BONDING AGENTS

Silane bonding agents contain two or more functional groups that each of them will interact with various type of molecules. These bonding agents have potential application in surface improvement of insulating layer. According to literature, the addition of silane bonding agents can increase the modulus and strength of propellant [10]. For example, in the study on the effects of silane bonding agent on the improvement of thermal stability and mechanical properties of HTPB/HMX composite system by [19], it was demonstrated that the addition of amino-silane in the HTPB/HMX matrix improved the mechanical properties of composite. Moreover, enhanced moisture resistance and thermal stability of composite are obtained. They demonstrated that the silane bonding agents, due to their considerable effect on tensile strength enhancement, are excellent candidates for application in HTPB/HMX composite systems. As a result, it can be concluded that the HTPB/HMX composite with silane bonding agent has higher mechanical properties than the ones with without silane bonding agent [19].

7.3.3 HYDANTOINS BONDING AGENTS

Today, hydantoins are known to be an important bonding agent for composite solid propellants containing nitramine. Hydantoin bonding agent could bring strong interaction between the binder and filler through covering particles as a sticky beads. DHE bonding agent, as a most popular hydantoin bonding agent, is suitable for linking oxidants (including RDX, HMX) in composite solid propellants. Using hydantoin bonding agent in composite solid propellant would make it easier to casting, molding, extruding, or mixing processes [18, 21]. The effect of DHE percent on the mechanical properties of composite solid propellants is illustrated in Table (7.1). It can be noted that the samples with 0.25 per cent DHE bonding agent have better mechanical properties than the other samples.

TABLE 7.1 Effect of DHE percent on the mechanical properties of composite solid propellants [21]

DHE%	293K (20°C)					223K (−50°C)		
	σ_m (MPa)	ε_m (%)	M (MPa)	ε_e (%)	$\varepsilon_{m\times}\sigma_m$ (MPa)	σ_m (MPa)	ε_m (%)	ε_e (%)
0.05	0.63	20	4.9	13	8.2	1.6	13	7.4
0.15	0.87	36	5.8	15	13	1.9	18	4.8
0.25	0.74	50	2.9	26	19	1.8	30	8.9

7.4 CONCLUSIONS

This chapter summarizes the effect of bonding agent on mechanical properties of composite solid propellants containing nitramine. Bonding agents can improve the mechanical properties of propellants by producing chemical or physical bond between binder and oxidizer. Among several bonding agent described earlier, neutral polymeric bonding agent (NBPA) is the most effective bonding agent that used in composite solid propellants containing nitramine. For example, the results show that the addition of 0.2 wt. per cent of NBPA bonding agent to HMX/PEG, results in higher elongation and mechanical properties.

KEYWORDS

- **Bonding agents**
- **Composite solid propellants**
- **Mechanical properties**
- **Nitramine propellants**

REFERENCES

1. Agrawal, J. P.; High Energy Materials Propellants, Explosives and Pyrotechnics. Weinheim: WILEY-VCH Verlag GmbH & Co. KGaA; **2010**.

2. Allen, H. C.; Bonding Agent for Nitramines in Rocket propellants. *US Patent* 4389263, **1983**.

3. Chung, H. L.; Kawata, K.; Itabashi, M.; Tensile strain rate effect in mechanical properties of dummy HTPB propellants. *J. Appl. Polym. Sci.* **1993**, *50*, 57–66.

4. Uscumlic, G. S.; Zreigh, M. M.; Dusan, Z. M.; Investigation of the interfacial bonding in composite propellants.1,3,5-Trisubstituted isocyanurates as universal bonding agents, *J. Serb. Chem. Soc.* **2006**, *71*, 445–458.

5. Behera, S.; Effect of RDX on elongation properties of AP/HTPB based case bonded composite propellants. *Sci. Spectrum,* **2009**, 31–36.

6. Li, X.; Jiao, J.; Yao, J.; Wang, L.; Study on the relationship between microscopic structure and mechanical properties of HTPB propellant. *Adv. Mater. Res.* **2011**, *4*, 1151–1155.

7. Duterque, G.; Lengellej, G.; Combustion mechanisms of nitramine-based propellants with additives. *J. Propul. Power,* **1990**, *6*, 718–725.

8. Shokri, S.; Sahafian, A.; Afshani, M. E.; Bonding agents and their performance mechanisms in composite solid propellants. MSc Thesis, K.N. Toosi University of Technology, Iran, **2005**.

9. Davenas, A.; Solid Rocket Propulsion Technology. Oxford: Pergamon Press; **1993**.

10. Kim, C. S.; Youn, C. H.; Nobel, P. N.; Gao, A.; Development of neutral polymeric bonding agents for propellants with polar composites filled with organic nitramine crystals. *Propell. Explos. Pyrot.* **1992**, *17*, 38–42.

11. Kim, C.; Sue, N.; Paul, N.; Youn, C. H.; Tarrant, D.; Gao, A.; The mechanism of filler reinforcement from addition of neutral polymeric bonding agents to energetic polar propellants. *Propell. Explos. Pyrot.* **1992**, *5*, 51–58.

12. Hamilton, R. S.; Wardle, R.; Hinshaw, J.; Oxazoline Bonding Agents in Composite Propellants. US Patent *5366572,* **1994**.

13. Ducote, M. E.; Carver, J. G.; Amine Salts as Bonding Agents. *US Patent* 4493741, **1987**.

14. Wallace, I. A.; Ambient Temperature Mix, Cast, and Cure Composite propellant Formulations. *US Patent* 5472532, **1995**.

15. Allen, H. C.; Clarke, F.; Tetra-Alkyl Titanates as Bonding Agents for Thermoplastic Propellants. *US Patent* 4597924, **1986**.

16. Hamilton, R. S.; Wardle, R.B.; Hinshaw, J. C.; Vinyl Ethers as Nonammonia Producing Bonding Agents In Composite Propellant Formulations. *US Patent* 5336343, **1994**.

17. Hasegawa, K.; Takizuka, M.; Fukuda, T.; Bonding Agents for AP and Nitramine/HTPB Composite Propellants. AIAA/SAE/ASME, 19th Joint Propulsion Conference, Washington, **1983**.

18. Consaga, J. P.; Dimethyl Hydantoin Bonding Agents in Solid Propellants. *US Patent* 4,214,928; **1980**.

19. Leu, A. L.; Shen S. M.: The effects of silane bonding agent on the improvement of thermal stability and mechanical properties of HTPB/HMX composite system. *Technol. Polym. Compd. Energ. Mater.* **1990**, 101–1.

20. Kim, C. S.; Filler reinforcement of polyurethane binder using a Neutral Polymeric Bonding Agent. *US Patent* 4915755, **1990**.

21. Perrault, G.; Lavertu, R.; Drolet, J. F; High-Energy Explosives or Propellant Composition. *US Patent* 4289551, **1981**.

CHAPTER 8

A STUDY ON ADSORPTION OF METHANE ON ZEOLITE 13X AT VARIOUS PRESSURES AND TEMPERATURES

FARSHID BASIRI, ALIREZA ESLAMI, MAZIYAR SHARIFZADEH, and MAHDI HASANZADEH

CONTENTS

8.1 INTRODUCTION

Natural gas, as a cleanest burning fossil fuel, has low carbon content and negligible sulfur dioxide emissions. Therefore, it is an excellent candidate for applications in diverse fields, and especially in heating and cooking on a domestic level and to power generation [1]. The use of natural gas is expected to increase in the utility, transportation, and industrial sectors in the near future [1, 2]. In general, natural gas is primarily methane, however, depending on the source and geographical location of production; it can contain large proportions of other compounds such as carbon dioxide, nitrogen, and small amounts of higher molecular weight hydrocarbons [3].

It is necessary to develop technologies that will allow us to utilize the fossil fuels while reducing the emissions of greenhouse gases [4]. There are many approaches to separate methane from air, such as cryogenic distillation of liquefied air and adsorption processes [5]. In recent years, separation and purification of gas mixtures by adsorption has been extensively studied in chemical and petrochemical industries [6, 7]. Pressure swing adsorption (PSA) attracts a growing interest due to its low energy requirements and low capital investment costs [7, 8]. This process was used to separate some gas species from a mixture of gases under pressure. It operates according to the molecular characteristics of gas species and affinity for an adsorbent material. In general, the higher the pressure, the more the gas being adsorbed. Special adsorptive materials (e.g., zeolites) are used as a molecular sieve, preferentially adsorbing the target gas species at high pressure [9, 10].

There have been various approaches to develop mathematical models for PSA process according to the literature. Several traditional adsorption models have been used by researchers, including Langmuir, Sips, Toth, UNILAN, and Dubinin-Astakhov [4]. In recent years, many algorithms grow out of nature—such as particle swarm optimization (PSO), differential evolution (DE) and, more recently, the cuckoo search (CS)—have been developed as an optimization tool for researchers. CS is a metaheuristic search algorithm that was inspired by the obligate parasitism of some cuckoo species by laying their eggs in the nests of other host birds [11].

In this chapter, adsorption of methane on zeolite 13X at various pressures and temperatures by PSA process was studied. Several adsorption isotherm models including Langmuir, Toth, Sips, UNILAN, and Dubinin-Astakhov were used to model the adsorption process of methane on zeolite

13X. The parameters of these adsorption models were optimized using CS algorithm.

8.2   EXPERIMENTAL

In this work, the data obtained from high-pressure adsorption of methane, on zeolite 13X were used to modeling this process. The working data of this research were obtained from the results of [12]. The data were modeled by the Langmuir, Toth, Sips, UNILAN, and Dubinin-Astakhov isotherms. The models and the parameters used in the fits are the same and are tabulated in Table 8.1. Fitting of the adsorption isotherm equations to experimental data was done with MATLAB mathematical software (version 7.6). The best fitting parameters were determined based on the minimum value of objective function, square of residuals, F_{obj}[12]:

$$F_{obj} = \sum_{T=1}^{a} \sum_{P=0}^{P_{max}} \sum_{S=1}^{n} (C_{exp} - C_{cal})^2 \tag{8.1}$$

where C_{exp} and C_{cal} are the methane concentrations, experimental and calculated, respectively,

T_i are the three different temperatures used, P_{max} is the maximum pressure of each isotherm, and S is the number of point per isotherm per gas.

TABLE 8.1   Traditional adsorption models that have been used to fit adsorption data for zeolite 13X

Isotherm	Model	Adjustable Parameter
Langmuir	$C_i = C_{mi} \dfrac{KP}{KP+1}$	C_{mi}, K
Toth	$C_i = C_{mi} \dfrac{(KP)^{1/n}}{1+(KP)^{1/n}}$	C_{mi}, K, n
Sips	$C_i = C_{mi} \dfrac{KP}{(1+(KP)^n)^{1/n}}$	C_{mi}, K, n

TABLE 8.1 *(Continued)*

Isotherm	Model	Adjustable Parameter
Unilan	$C_i = C_{mi} \dfrac{1}{2n} Ln \left[\dfrac{1 + KP_0 \exp(n)}{1 + KP_0 \exp(-n)} \right]$	C_{mi}, K, n
Dubinin-Astakhov	$C_i = C_{mi} \exp \left\{ - \left[\dfrac{RT}{E} \ln \left(\dfrac{P^{sat}}{P} \right) \right]^n \right\}$	C_{mi}, E, n

8.3 RESULTS AND DISCUSSION

The adsorption isotherms of methane on zeolite 13X at 298, 308, and 323 K are shown in Figure 8.1, and the parameters of the fitting are reported in Table 8.2. As illustrated in this the figure, the data were well fitted with all models. Both the Toth and Dubinin-Astakhov models have a very good flexibility to fit experimental data. However, Dubinin-Astakhov model shows much lower objective function (F_{obj}) than the Toth model. Adsorption equilibrium of methane has been compared with previous literature [4, 7] and showed good agreement with previously published data. Comparing the results obtained for this system to those obtained by [4] shows that we get a much improved objective function by usign CS algorithm. The improvement of both the Toth and Dubinin-Astakhov models was 0.15 and 1.47, respectively.

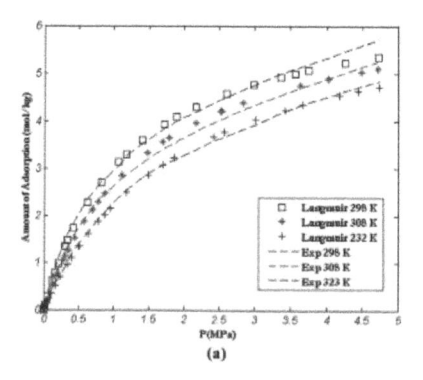

FIGURE 8.1 *(Continued)*

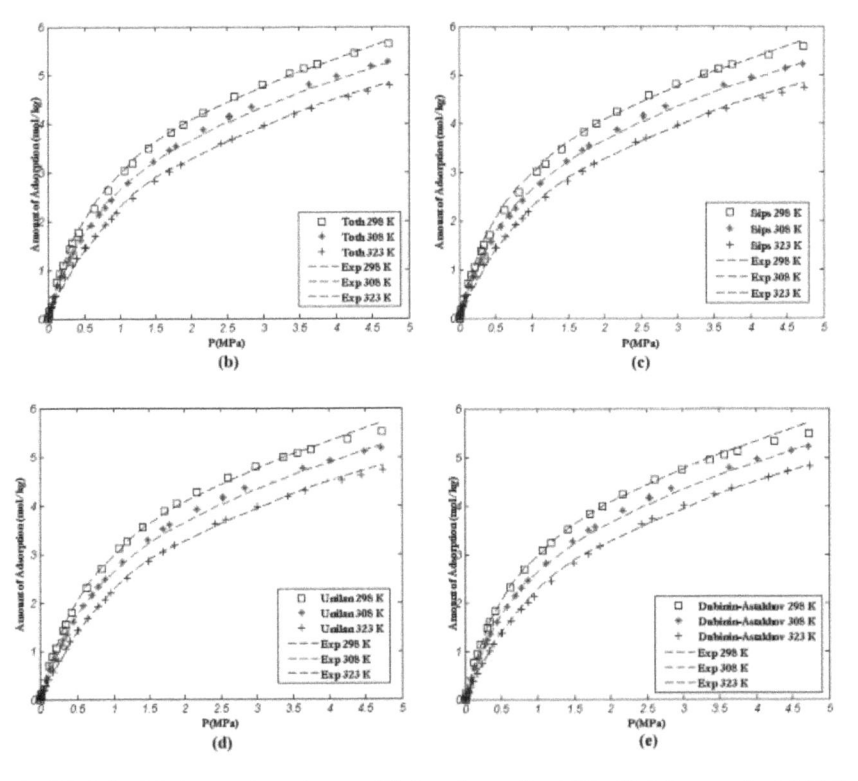

FIGURE 8.1 Methane adsorption equilibrium in zeolite 13X of (a) Langmuir, (b) Toth, (c) Sips, (d) Unilan, and (e) Dubinin-Astakhov models.

TABLE 8.2 Optimum values for the fitting of the models

Isotherm	C_{mi} (mol/kg^{-1})	n	H_{st} (kj/mol^{-1})	K_0 (Mpa^{-1})	E (j/mol^{-1})	F_{obj}
Longmuir	6.73	-	15.95	0.0013	-	0.76
Toth	11.99	0.526	15.33	0.0017	-	0.24
Sips	8.40	1.13	12.498	0.0036	-	0.32
UNILAN	7.45	1.34	15.52	0.0012	-	0.40
Dubinin-Astakhov	7.36	1.87	-	-	0.0088	0.23

8.4 CONCLUSIONS

This chapter presented a study on the adsorption equilibrium isotherms of methane on zeolite 13X at high-pressure process. The data were well fitted with the Langmuir, Toth, Sips, UNILAN, and Dubinin-Astakhov models according to the cuckoo search algorithm. Although all models have a very good flexibility to fit experimental data, Dubinin-Astakhov model shows much lower objective function (F_{obj}) than the other models. The data presented compare very well with previously reported values.

ACKNOWLEDGMENTS

The authors thank S. Cavenati, C. A. Grande, and A. E. Rodrigues at University of Porto for their data.

KEYWORDS

- **Adsorption models**
- **Cuckoo search algorithm**
- **Methane**
- **Pressure swing adsorption**

REFERENCES

1. Blaha, D.; Bartlett, K.; Czepiel, P.; Harriss, R.; Crill, P.; Natural and anthropogenic methane sources in New England, *Atmos. Environ.* **1999**, *33*, 243–255.
2. Hui, K. S.; Chao, C. Y. H.; Kwong, C. W.; Wan, M. P.; Use of multitransition-metalion-exchanged zeolite 13X catalysts in methane emissions abatement, *Combust. Flame.* **2008**, *153*, 119–129.
3. Wang, Y.; Hashim, M.; Ercan, C.; Khawajah, A.; Othman, R.; High Pressure Methane Adsorption on Granular Activated Carbons, 21st Annual Saudi-Japan Symposium, November 2011, Catalysts in Petroleum Refining & Petrochemicals, Dhahran, Saudi Arabia; **2011**.
4. Behvandi, A.; Tourani, S.; Equilibrium modeling of carbon dioxide adsorption on zeolites. *World Acad. Sci. Eng. Technol.* **2011**, *76*.

5. Salil, U.; Rege, R.; Yang, T.; Qian, K.; Buzanowski, M. A.; Air-purification by pressure swing adsorption using single/layered beds. *Chem. Eng. Sci.* **2001**, *56*, 2745–2759.

6. Jee, J. G.; Kim, M. B.; Lee, C. H.; Adsorption characteristics of hydrogen mixtures in a layered bed: binary, ternary and five component mixtures. *Ind. Eng. Chem. Res.* **2001**, *40*, 868–878.

7. Cavenati, S.; Grande, C. A.; Rodrigues, A. E.; Adsorption equilibrium of methane, carbon dioxide, and nitrogenon zeolite 13X at high pressures. *J. Chem. Eng.* **2004**, *49*, 1095–1101.

8. Moghadaszadeha, Z.; Towfighib, J.; Mofarahi, M.; Four-bed pressure swing adsorption for oxygen separation from air with zeolite 13X, iran international zeolite conference (IIZC'08), *29* April 2008, Tehran, Iran; **2008**.

9. Mofarah, M.; Sadrameli, M.; Towfighi, J.; Four-bed vacuum pressure swing adsorption process for propylene/propane separation. *Ind. Eng. Chem.* **2005**, *44*, 1557–1564.

10. Yoshida, S.; Ogawa, N.; Kamioka, K.; Hirano, S.; Mori, T.; The study of zeolite molecular sieves for production of oxygen by using presuure swing adsorption. *Kluwer Acad. Pub.* **1999**, *5*, 57–61.

11. Subotic, M.; Tuba, M.; Bacanin, N.; Simian, D.; Parallelized cuckoo search algorithm for unconstrained optimization. *World Sci. Eng. Acad. Soc.* **2012**.

12. Santos, J.C.; Cruz, P.; Regala, T.; Magalhaes, F. D.; Mendes, A.; High-purity oxygen production by pressure swing adsorption. *Ind. Eng. Chem.* **2007**, *46*, 591–599.

CHAPTER 9

IMPORTANCE OF THE PHASE BEHAVIOR IN BIOPOLYMER MIXTURES

Y. A. ANTONOV and PAULA MOLDENAERS

CONTENTS

9.1   INTRODUCTION

The importance of the phase behavior in biopolymer mixtures is evident in many technological processes, such as isolation and fractionation of proteins (see, e.g., Refs. [1–6]), and enzymes [7], enzyme immobilization [8–9], encapsulation [10], and drug delivery [9, 11]. Aqueous, two-phase systems are used in modern technological processes where clarification, concentration, and partial purification are integrated in one step [12]. Thermodynamic incompatibility, or, in other words, segregative phase separation determines the structure and physical properties of biopolymers mixtures in quiescent state [13–15] and under flow [16–18] and plays an important role in protein processing in food products [14].

From a technological viewpoint, especially important are biopolymer systems that undergo liquid–liquid phase separation in a wide concentration range, starting from low concentrations [19]. But whether phase separation is desired or not, it is important for practical applications to understand the underlying mechanisms and molecular interactions that govern the phase behavior of a given system [20]. Despite the considerable amount of research in the field of segregating polymer mixtures, the molecular interactions in the systems are inadequately understood, although theoretical models have been proposed [21–28]. There have, as of yet, been comparatively few studies on phase separation in mixtures of similarly charged polyelectrolytes [29, 30]. Such systems may have advantages over uncharged systems in the separation of proteins due to the tunable charge in the system arising from the dissociated counter ions of the polyelectrolytes [29, 30]. Although the majority of biopolymer mixtures show phase separation [14, 32], in most cases the phase separation takes places at critical total concentrations, which are much higher (7–12 wt%) [31, 32] compared with those of synthetic polymers (less than 1–2 wt%). Unlike synthetic polymers with flexible chains, many proteins are known to be relatively symmetric compact molecules and are usually able to form solutions that can still be considered dilute for concentrations tenfold higher than for synthetic polymers of the same molecular weight [33].

The aims of this study is to induce demixing in semidilute and highly compatible sodium caseinate/sodium alginate system (SC–SA) mixtures in the presence of sodium salt of dextran sulfate (DSS) at pH 7.0, (above the isoelectric point of caseins), and to characterize phase equilibrium, intermacromolecular interactions, and structure of such systems by rheo-small angle light scattering (SALS), optical microscopy (OM), phase analysis, dynamic light scattering (DLS), fast protein liquid chromatography (FPLC), ESEM, and rheology. The molecular weight, charge, and topography of the accessible surface of water-soluble complexes of proteins with anionic polysaccharides differ markedly from the "free" proteins. Therefore, it can be assumed that all these factors may affect the phase separation. In the present work, we focus our study on the phase transitions in aqueous semidilute homogeneous sodium caseinate/sodium alginate (SC–SA) systems with the total concentration of biopolymers 1, 5 wt per cent–2.5 wt per cent, that is, much below the critical concentrations for phase separation [17]. The phase state of the SC–SA mixtures is not sensitive to changes in pH, ionic strength, and temperature in the quiescent state [31, 32] and under of shear flow [17]. Therefore, the effect of demixing that can be reached for this system can be easily reproduced for other emulsions in which the phase equilibrium is more sensitive to physicochemical parameters. Here, it will be explored how far this strategy of demixing can be extended to other biopolymer pairs. For this reason, gelatin SA and gelatin SC systems will be investigated to assess the generality of our observations. In addition, the shear-induced behavior of the decompatibilized semidilute SC–SA system will be presented and compared with that of the "native" SC–SA system.

Alginate is an anionic polysaccharide consisting of linear chains of (1–4)-linked ß-D-mannuronic and α-L-guluronic acid residues. These residues are arranged in blocks of mannuronic or guluronic acid residues linked by blocks in which the sequence of the two acid residues is predominantly alternating [33, 34]. Casein is a protein composed of a heterogeneous group of phosphoproteins organized in micelles. These biopolymers are well known, widely used in industry for their textural and structuring properties [14, 31, 32, 33, 35], and the thermodynamic behavior of the ternary water–caseinate–alginate systems is known from literature [17, 31, 32, 35].

9.2 MATERIALS AND METHODS

The caseinate at neutral pH is negatively charged, such as alginate and DSS. The sodium caseinate sample (90% protein, 5.5% water content,

3.8% ash, and 0.02% calcium) was purchased from Sigma Chemical Co. The isoelectric point is around pH = 4.7–5.2 [36]. The weight average molecular mass of the sodium caseinate in 0.15 M NaCl solutions is 320 kDa. The medium viscosity sodium alginate, extracted from brown seaweed (*Macrocystis pirifera*), was purchased from sigma. The weight average molecular weight of the sample, M_w was 390 kDa [16]. Dextran sulfate, DSS (M_W = 500 kDa, M_n = 166 kDa, η (in 0.01 M NaCl) = 50 mL/g, 17% sulfate content, free SO_4 less than 0.5%) was produced by Fluka, Sweden (Reg. No. 61708061 A, Lot No. 438892/1). The gelatin sample used is an ossein gelatin type A 200 Bloom produced by SBW Biosystems, France. The Bloom number, weight average molecular mass, and the isoelectric point of the sample reported by the manufacturer are 207, 99.3 kDa, and 8–9, respectively.

Preparation of the protein/polysaccharide mixtures: Most experiments were performed in the much diluted phosphate buffer (ionic strength, I = 0.002). To prepare molecularly dispersed solutions of SC, SA, gelatin, or DSS with the required concentrations, phosphate buffer (Na_2HPO_4/NaH$_2PO_4$, pH 7.0, I = 0.002, and 0.015) was gradually added to the weighed amount of biopolymer sample at 298 K, and stirred, first for 1h at this temperature and then for 1h at 318K. The solutions of SC, SA, and DSS were then cooled to 296K and stirred again for 1h. The required pH value (7.0) was adjusted by addition of 0.1–0.5M NaOH or HCl. The resulting solutions were centrifuged at 60,000g for 1h at 296K or 313K (gelatin solutions) to remove insoluble particles. Concentrations of the solutions are determined by drying at 373K up to constant weight. The ternary water–SC–SA systems with required compositions were prepared by mixing solutions of each biopolymer at 296K. After mixing for 1 h, the systems were centrifuged at 60,000 *g* for 1 h at 296K to separate the phases using a temperature-controlled rotor.

Determination of the phase diagram: The effect of the presence of DSS on the isothermal phase diagrams of the SC–SA system was investigated using a methodology described elsewhere [36]. The procedure is adapted from Koningsveld and Staverman [37] and [38]. The weight DSS/SC ratio in the system, (q) was kept at 0.14. The threshold point was determined from the plot as the point where the line with the slope−1 is tangent to the binodal. The critical point of the system was defined as the point where the binodal intersects the rectilinear diameter, which is the line joining the centre of the tie lines.

Rheo-optical study: A rheo-optical methodology based on small angle light scattering (SALS) during flow, is applied to study *in situ* and on a time-resolved basis the structure evolution. Light scattering experiments were conducted using a Linkam CSS450 flow cell with a parallel-plate geometry. A5 mW He–Ne laser (wavelength 633 nm) was used as light source. The 2D scattering patterns were collected on a screen by semi-transparent paper with a beam stop and recorded with a 10-bit progressive scan digital camera (Pulnix TM-1300). Images were stored on a computer with the help of a digital frame grabber (Coreco Tci-Digital SE). The optical acquisition set-up has been validated for scattering angles up to 18°. The gap between the plates has been set at 1 mm, and the temperature was kept constant by means of a thermostatised water bath. In-house developed software was used to obtain intensity profiles and contour plots of the images (New SALS SOFTWARE-K.U.L.). Turbidity measurements have been performed by means of a photodiode.

Microscopy observations during flow have been performed on a Linkham shearing cell mounted on a Leitz Laborlux 12 Pols optical microscope using different magnifications.

Rheological measurements were performed using a Physica Rheometer, type CSL2500 A/G H/R, with a cone-plate geometry CP50-1/Ti ~ diameter 5 cm, angle 0,993°, Anton Paar. The temperature was controlled at 23°C by using a Peltier element. For each sample, flow curves were measured at increasing shear rate ~ from 0.1 to 150s^{-1}. The ramp mode was logarithmic and the time between two measurements was 30s. Frequency sweeps ~0.1–200 rad/s were carried out as well for a strain of 3.0 per cent, which was in the linear response regime. During the rheological measurements, all samples were covered with paraffin oil to avoid drying.

Dynamic light scattering: Determination of intensity-weighted distribution of hydrodynamic radii (R_H) of SC, SA, and DSS solutions and their mixtures was performed, using the Malvern ALV/CGS-3 goniometer. Concentration of the protein in protein-dextran sulfate mixtures was kept at 0.1 (w/w). For each sample, the measurement was repeated three times. The samples were filtered before measurement through DISMIC-25cs (cellulose acetate) filters (sizes hole of 0.22 μm for the binary water-casein and water-dextran sulfate solutions and 0.80 μm for the protein–polysaccharide mixtures). Subsequently, the samples were centrifuged for 30 sec at 4,000 g to remove air bubbles, and placed in the cuvette housing that was kept at 23°C in a toluene bath. The detected scattering light intensity was processed by digital ALV-5000 correlator software. The second-order cumulant fit was used for the determination of the hydrodynamic radii. The asymmetry coefficient (Z) of the complex particles was estimated by Debye method based on determining the scattering intensity at two angles 45° and 135°, symmetrical to the angle 90°.

Zeta potential measurement: The ζ potential measurements of SC and DSS solutions and their mixtures at different q values were performed at 23°C with a Malvern-Zetamaster S, model ZEM 5002 (England), using a rectangular quartz capillary cell. The concentration of the protein in solutions was 0.1 wt per cent, and the concentrations of DSS in the protein-polysaccharide solutions were variable. All solutions were prepared in phosphate buffer (Na_2HPO_4/NaH_2PO_4, pH 7.0, $I = 0.002$). The zeta potential was determined at least three times for each sample. The zeta potential was calculated automatically from the measured electrophoretic mobility, by using the Henry equation:

$$U_e = \varepsilon z \rho f / 6\pi \eta, \tag{9.1}$$

where U_e is electrophoretic mobility, ε is the dielectric constant, η is the viscosity, and $z\rho$ is the zeta potential. The Smoluchowski factor, $f = 1.5$ was used for the conversion of mobility into zeta potential.

Environment scanning electron microscopy: Microstructural investigation was performed with the environment scanning electron microscope Philips XL30 ESEM FEG. The instrument has the performance of a conventional SEM but has the additional advantage that practically any material can be examined in its natural state. The samples were freeze-fractured in freon and immediately placed in the environment scanning electron microscope (ESEM). Relative humidity in the ESEM chamber (100%) was maintained using a Peltier stage. Such conditions were applied to minimize solvent loss and condensation, and control etching of the sample. Images were obtained within less than 5 min of the sample reaching the chamber. The ESEM images were recorded multiple times and on multiple samples to ensure reproducibility.

Fast protein liquid chromatography or FPLC. Solutions of sodium caseinate, (0.5 wt%), dextran sulfate (0.5 wt%) and their mixtures, containing 0.5 wt per cent of the protein and variable amount of dextran sulfate were applied on a Superose 6 column (HR 10/30), Amersham Biosciences mounted on an FPLC apparatus (Pharmacia, Uppsala, Sweden). Elution was performed at room temperature with phosphate buffer (5mM Na_2H-PO_4/NaH_2PO_4, pH 7.0) 2 per cent (v/v) n-propanol (Riedel-de Haen, Seelze, Germany) and 0.015 M NaCI. The samples and the elution buffer were filtered through a 0.22 um sterile filter. The flow rate was 0.2 mL min-1 and the column was monitored by UV detection at 214 nm.

Determination of dextran sulfate content: The phenol-sufuric acid method of [39]., was applied. 50 uL. 80 per cent (w/w) phenol in water and 5 mL sulfuric acid were added to the measured samples of 0.5 mL. After 30 min at room temperature, the absorbance at 485 nm was measured. A calibration plot was constructed with D-glucose (Riedel-de Haen).

9.3   RESULTS AND DISCUSSION

9.3.1   DSS-INDUCED DEMIXING

The experimental results shown in this section have been obtained on water (97.5 wt%)-SC (2.00 wt%)-SA (0.5 wt%) semidilute systems. This system is located in the one-phase region far from the binodal line. To study the effect DSS on the phase behavior, a flow history consisting of two shear zones is used. First, a preshear of 0.5 s^{-1} is applied for 1,000 s (500 strain units) to ensure a reproducible initial morphology. Subsequently, this preshear is stopped, and the sample is allowed to relax for 30 sec leaving enough time for full relaxation of deformed droplets. Then, SALS patterns are monitored.

The SALS patterns and the scattering intensity upon adding different amount of DSS are shown in Figures 9.1 (a–f), and 9.2, starting from a concentration of DSS as low as 2.08 10^{-3} wt per cent. In the absence of DSS, no scattered light is observed (data are not presented). The presence of even only 2.08 10^{-3} wt per cent DSS in the homogeneous system led to appreciable increase the SALS pattern (Figure 9.1), and accordingly the light scattering intensity (Figure 9.2). It is important to note that that the SC-DSS system remains homogeneous in the DSS concentration range studied here. Centrifugation of the SC-DSS systems (120 min, 60.000 g, 296 K) prepared at the same conditions did not show phase separation.

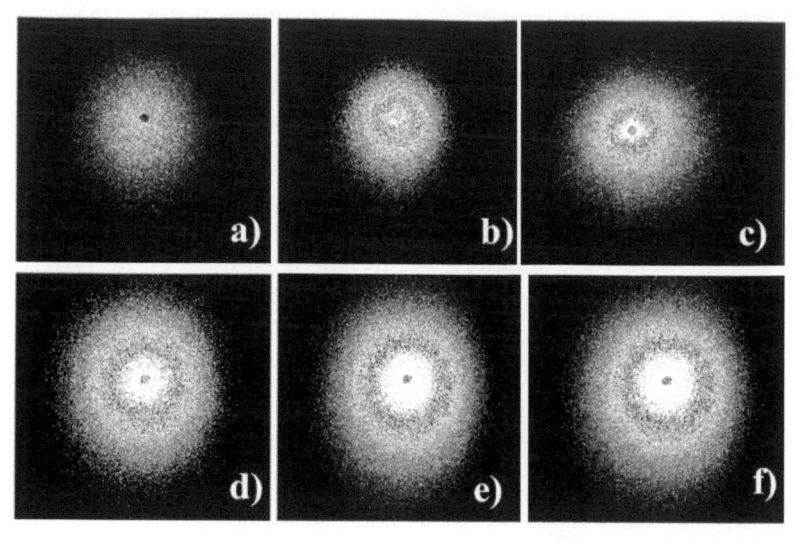

FIGURE 9.1 Effect of the concentration of DSS on the SALS patterns of water (97.5 wt%)-SC (2.00 wt%)-SA (0.5 wt%) single-phase systems. pH 7.0. $I = 0.002$ (phosphate buffer); temperature 296 K; and concentrations of DSS in mixture, wt per cent: (a) 2.08×10^{-3}, (b) 4.10×10^{-3}, (c) 1.61×10^{-2}, (d) 7.50×10^{-2}, (e) 0.15, (f) 0.29, and resulting DSS/ SC ratio: (a) 0.001, (b) 0.002, (c) 0.008, (d) 0.0375, (e) 0.075, and (f) 0.145.

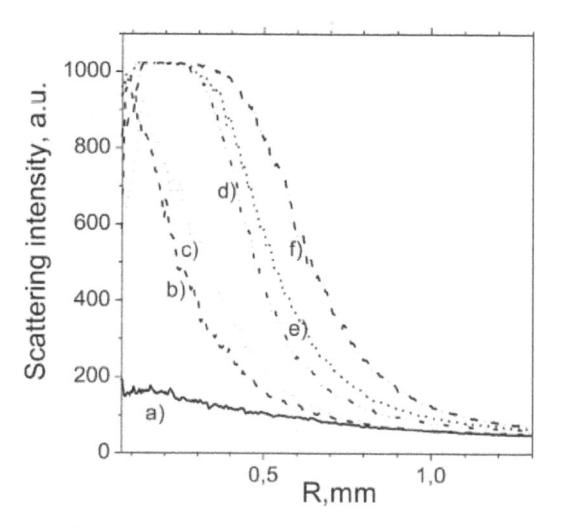

FIGURE 9.2 Effect of the concentration of DSS on the scattering intensity of water (97.5 wt%)-SC (2.00 wt%)-SA (0.5 wt%) single-phase systems as a function of the distance from the bean stop. The other parameters are the same as in Figure 9.1.

When the DSS concentration in the SC–SA system increases, the SALS pattern (Figure 9.1) and the scattering intensity (Figure 9.2) of the system sharply grows. This indicates that the position of the system on the phase diagram changes deeply into the two phase range. The corresponding microscopy images for the same concentrations of DSS and the same flow conditions are shown in Figure 9.3. One can see that the phase separation led to formation of liquid–liquid emulsions. At the lowest DSS concentration (2.08 10-3 wt%), the system contains ultrasmall droplets of the dispersed phase having a size of 2–3 μm. At higher DSS concentrations, the size of the droplets increases significantly in agreement with SALS data achieving more than 50 μm in diameter.

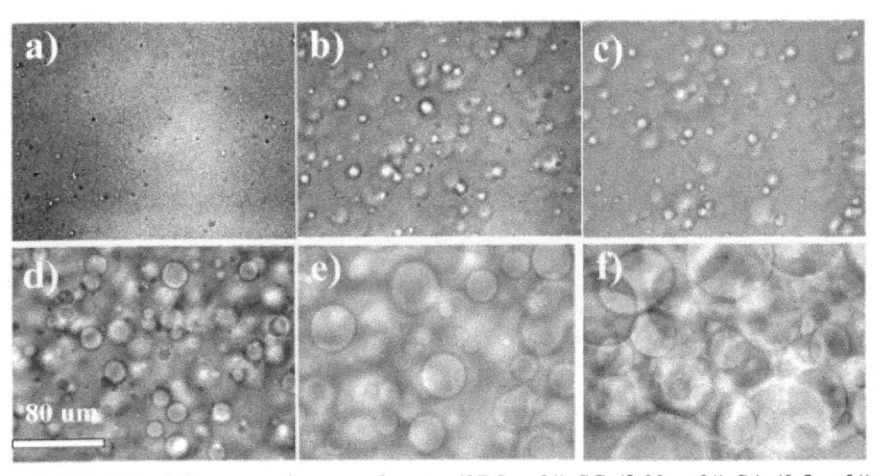

FIGURE 9.3 Microscopy images of water (97.5 wt%)-SC (2.00 wt%)-SA (0.5 wt%) system after addition of different amounts of DSS. pH 7.0, $I = 0.002$ (phosphate buffer). temperature 296 K. The other parameters are the same as in Figure 9.1.

To quantify the effect of DSS on phase equilibrium in semi-diluted SC–SA system, the isothermal phase diagram of the system was determined in the presence of DSS, at DSS/SC weight ratio (q) = 0.14, plotted in the classical triangular representation, and compared with that obtained in the absence of DSS (Figure 9.4). The phase separation in the presence of DSS has a segregative character with preferential concentrating of SC and SA in different phases. The phase diagram of the initial system, without DSS, is characterized by a high total concentration of biopolymers at the critical point ($C_c^t = 62.9$ g/L), and a strong asymmetry ($Ks = 15.5$).

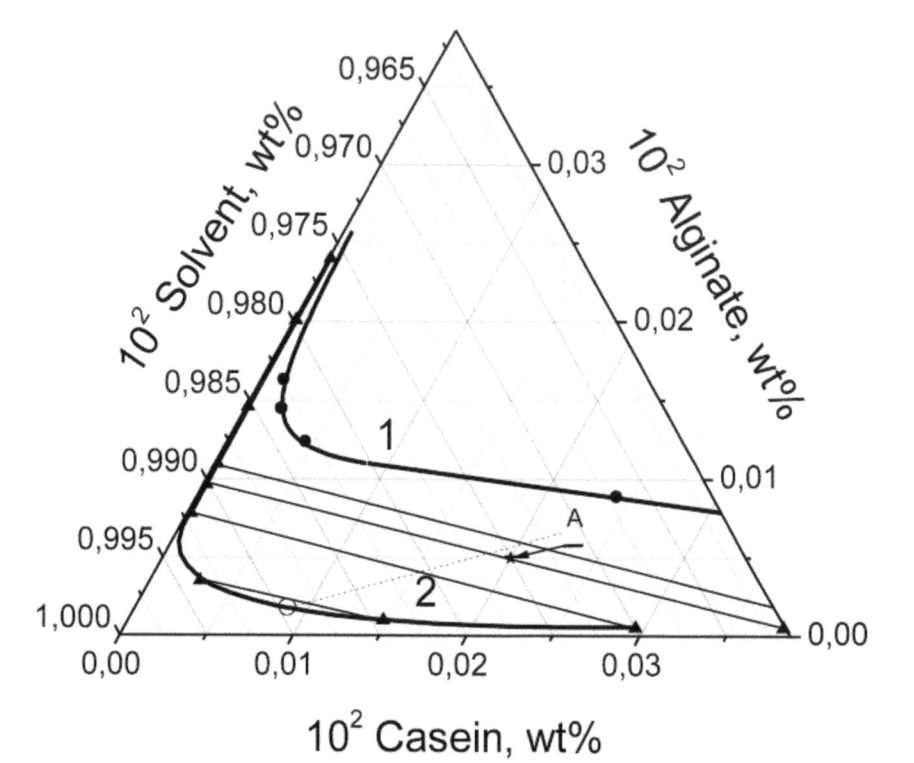

10^2 Casein, wt%

FIGURE 9.4 Isothermal phase diagrams of the W–SC–SA system. pH 7.0, $I = 0.002$ (phosphate buffer), 296 K. 1, In the absence of DSS and 2, in the presence of DSS, at DSS/SC weight ratio $(q) = 0.14$.

The presence of DSS affects dramatically the phase separation, significantly increasing the concentration range corresponding to two phase state of the system. The total concentrations of biopolymers at the critical point decreases to 10.6 g/L. The phase separation is observed at total concentrations of biopolymers just above 1 wt per cent, i.e., level of compatibility of the biopolymers after an addition of DSS seems to be one of the smallest known for biopolymer mixtures (see, e.g., [40, 41]). The decrease in compatibility of casein and alginate is especially surprising when taking into account that the phase composition of this system is weakly dependent on many physicochemical factors, such as pH (in the pH range from 7 to 10), ionic strength and temperature (from 5 to 60°C) [17, 32, 34].

9.3.2 RHEOLOGICAL BEHAVIOR OF THE DEMIXED SYSTEMS

For the rheological investigations, the homogeneous W-SC (2.0 wt%)-SA (0.5 wt%) system (point A on the phase diagram, Figure 9.4) was characterized before, and after addition of DSS at the DSS/SC weight ratio, $q = 0.045$, and $q = 0.15$, respectively. The latter two systems were two-phase ones with the content of the casein enriched phase 15 w/w, and 55 per cent w/w accordingly. The experimental flow protocol applied was the same as the one used for rheo-SALS. The mechanical spectrum and flow curve were determined to characterize the state of the systems through their viscoelastic behaviors. It has been shown [42] that at moderately low-shear rates, the biopolymer emulsions can be regarded as conventional emulsions and various structural models that are available in the literature for prediction of the morphology in these emulsions can also be used for prediction of the structure in aqueous biopolymer emulsions.

The evolution of the mechanical spectrum was investigated as a function of DSS concentration. These viscoelastic behaviors were monitored and compared with the behavior of the W–SC–SA system without DSS. The dynamic modulus G' (elastic) and G'' (viscous) were measured with frequency sweep experiments at a constant strain of 3 per cent, which was checked as being in the linear regime. The obtained data are presented in Figure 9.5.

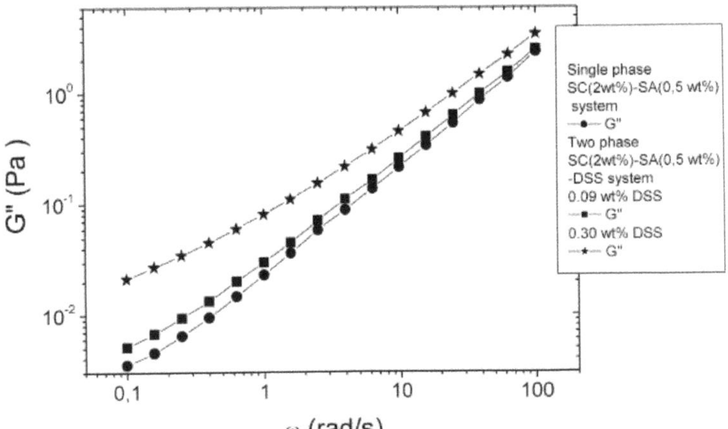

FIGURE 9.5 Dynamic spectra of single-phase W–CS (2 wt%)-SA (0.5 wt%) system, and two phase W–CS (2 wt%)-SA (0.5 wt%)-DSS systems. pH 7.0, $I = 0.002$ (phosphate buffer), 296 K.

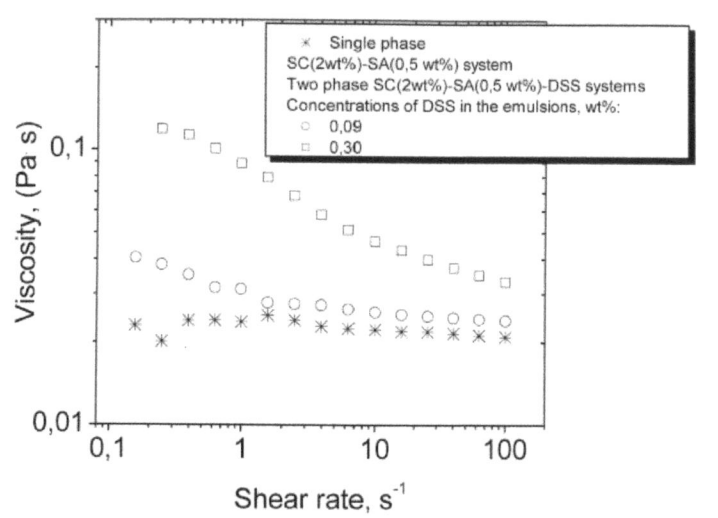

FIGURE 9.6   Flow viscosity of single-phase W–CS (2 wt%)-SA (0.5 wt%) system, and two-phase W–CS (2 wt%)-SA (0.5 wt%)-DSS Systems, after application of increasing shear rates. pH 7.0, $I = 0.002$ (phosphate buffer), 296 K.

For the single-phase system, and the system containing 0.09 wt per cent DSS, G' was too low to be measured accurately. Under these conditions, the system behaves as purely viscous liquid with the curve of G'' versus frequency displaying a slope of one on a double logarithmic graph. In the present of DSS, the system undergoes phase separation, and this transition leads to an appreciable increase of the moduli. The elastic properties of the decompatibilized W–SC–SA system were mainly induced by the presence of the DSS. In the presence of high ionic strength (0.25, NaCl), when electrostatic interactions were suppressed the mechanical spectrum of the system ($q = 0.14$) becomes insensitive to the presence of DSS (data are not presented). Flow curves determined at the same concentrations show an increase in viscosity for the demixed systems, especially remarkable at a low shear rates (Figure 9.6).

More detailed experiments were then carried out on the single phase W–SC (4 wt%)-DSS (variable), and W–SA (0.5 wt%)-DSS (variable) systems to understand how DSS affects the mechanical spectrum of the casein and alginate solutions, and accordingly the coexisting phases. The behavior of these solutions in the presence of sulfated polysaccharide is clearly different (Figures 9.7 and 9.8), the casein-enriched phase is sensitive to the presence of DSS, whereas the viscoelastic properties of the alginate-enriched phase in the presence of DSS remain almost unaltered. As reported in Figure 9.7a, the dependence of the G'' on the DSS/casein ratio has an extreme character, with a maximum at a DSS/casein ratio around 0.14. In the presence of even small amounts of DSS (0.01–0.05 wt%), a dramatic increase of the G'' of the emulsion takes place. Thus, in the presence of 0.5 wt per cent of DSS (at $q = 0.14$) and at a frequency 1 rad/s, G'' values is more than 1,400 times, higher compared with those of the single-phase system with almost the same composition. From theory, we know that such dependences are typical for the formation of interpolymer complexes [42]. Similar changes were observed for the viscosity (Figure 9.8 ab). At $q = 0.14$ and a shear rate of 10 s^{-1}, the viscosity is more than 940 times higher compared with those of the single-phase system with almost the same composition (Figure 8b). It is important to note that in the shear rate range from 0.1 to 150 s^{-1}, we did not find any difference in the flow curves obtained in conditions with increasing versus decreasing shear rate (data are not presented). It can be assumed that the dramatic changes in rheological behavior of the casein-alginate system in the presence of DSS are due to interactions of the casein molecules with the DSS molecule.

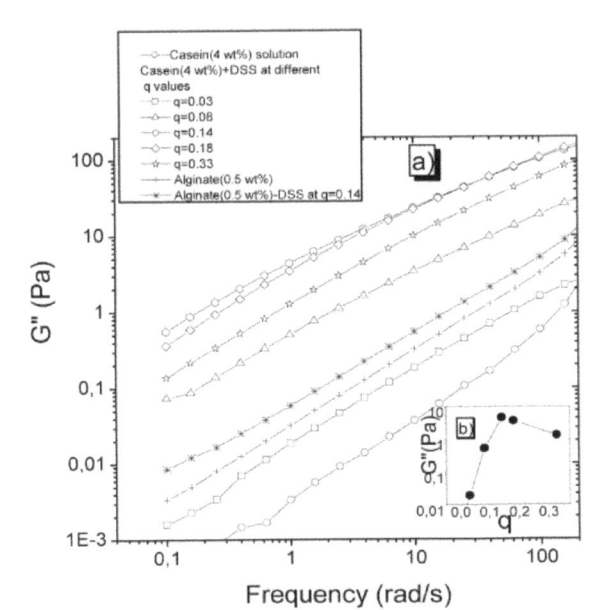

FIGURE 9.7 (a), G'' of W–SA(0.5 wt%), W–SA(0.5 wt%)-DSS, and W–SC(4 wt%)—DSS, systems at different q values, (b) the dependence of G'' on q values for W–SC(4 wt%)—DSS, system at frequency 1.0 rad/s. pH 7.0, $I = 0.002$ (phosphate buffer), 296 K.

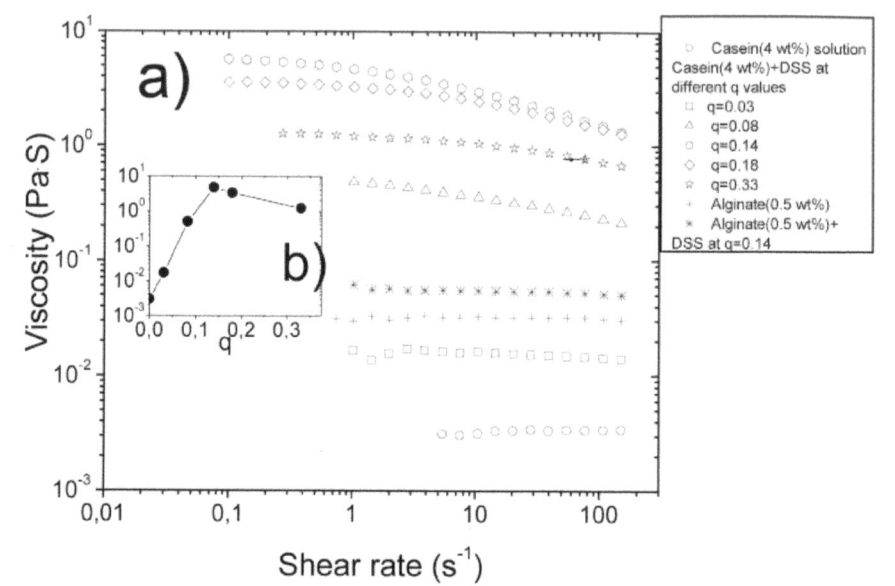

FIGURE 9.8 (a) Dependences of flow viscosity of W–SC(4 wt%), W–SA(0.5 wt%), and W–SC (4 wt%)—DSS (var) systems (b) and the dependence of flow viscosity of the W–SC–DSS system on q values at shear rate 1.0 s^{-1}. pH 7.0, $I = 0.002$ (phosphate buffer), 296 K.

It can be suggested that casein interacts with DSS, and this interaction may have an effect on the phase separation. Note that the viscosity of the demixed system in Figure 9.8 decreased from 5.72 to 1.74 Pa s with increasing shear rate from 0.1 to 150 sec^{-1}, which highlights the shear thinning behavior of the demixed system, indicating a structural change. The result was striking since most concentrated protein–polysaccharide mixtures can be shear thinning only due to the polysaccharide relaxations. In the absence of structure-induced formation, the rheological behavior of concentrated polysaccharide solutions is monotonically shear thinning; the viscosity varies between two extremes η_o and η_∞. A possible additional mechanism would be the breakdown of structures due to the breakup of physical bonds at high shear. This structure was most probably due to the electrostatic interactions between SC and DSS. Indeed, in the presence of 0.25 M NaCl, when no attractive interaction took place, no shear thinning behavior was observed (data are not shown). More detailed experiments were then carried out to understand the mechanism of demixing.

9.3.3 INTERMACROMOLECULAR INTERACTIONS AND THE MECHANISM OF DEMIXING IN SC–SA–DSS SYSTEM

An important property of the demixed semidilute SC–SA systems described above is their high stability against homogenization and low sensitivity to change in temperature. Thus, for the mixtures with different composition, we observed constancy of absorption values at 500 nm during 6h storage, as well as in processes of their heating from +5°C to 70°C. The results obtained (Figures 9.7 and 9.8) show the presence the intermacromolecular interactions between SC and DSS. Usually, coulomb protein–polysaccharide complexes are formed only in the vicinity of the isoelectric point of the protein [44]; but for several systems, formation of soluble protein—polysaccharide complexes has been registered even at pH 6–8.0 [45–47]. A beneficial consequence of complexation of sulfated polysaccharide with caseins at pH values above IEP is the protection afforded against loss of solubility as a result of protein aggregation during heating or following high-pressure treatment [48, 49].

The mechanism of this protection has been unclear until now. Snoeren, Payens, Jevnink and both, assumed [50] that there is a nonstatistical distribution of positively charged aminoacid residues along the polypeptide chain of kappa casein molecules and, as a consequence, the existence of a dipole interacting by its positive pole with sulfur polysaccharide is responsible for complex formation in such systems.

Many scientists suppose [51, 52] that nonelectrostatic forces, hydrophobic, and (or) hydrogen bonds play a determinant role in this process. In the case of sulfated polysaccharides, this assumption is confirmed by experimental data showing the capacity of the sulfate groups to form hydrogen bonds with the protein cationic groups [53].

Introduction of NaCl in the initial buffer results in full insensitivity of the viscosity and the phase diagram of the SC–SA system to the presence of DSS in all the q range studied. On the other hand, an addition of 0.2 M NaCl in the SC–SA–DSS system at $q = 0.14$ after a 24 h storage results in a sharp increase in the level of compatibility of SC with SA to that of SC–SA solution alone. This shows that the complexes are formed and stabilized via electrostatic interaction, rather than through hydrogen bonds formation or hydrophobic interaction. The role of salt is to "soften" the interactions, which is equivalent to making the electrostatic binding constant smaller.

To study intermacromolecular interactions in the process of demixing of the SC–SA system, at first, we focus our attention to the interaction between SC and DSS in aqueous solutions within the region of pair interaction. To this aim, we have chosen SC and DSS concentrations low enough to exclude or considerably diminish effects of possible aggregation. This allows us to single out information on interaction processes between the two types of macromolecules, well separated from the subsequent aggregation process. DLS can provide information about the hydrodynamic radius of proteins and polysaccharides and about the binding of ligands to these types of macromolecules. Figure 9.9 shows the intensity-weighted distribution of hydrodynamic radii (R_H) of solutions of sodium caseinate, dextran sulfate and their mixtures with the concentration of the protein equal to 0.1 (w/w), that is, at the total concentrations below the critical concentration of phase separation of SC–SA system (see Figure 9.4). At 296 K, molecules of SC and DSS have R_H values 119 nm and 250 nm, respectively.

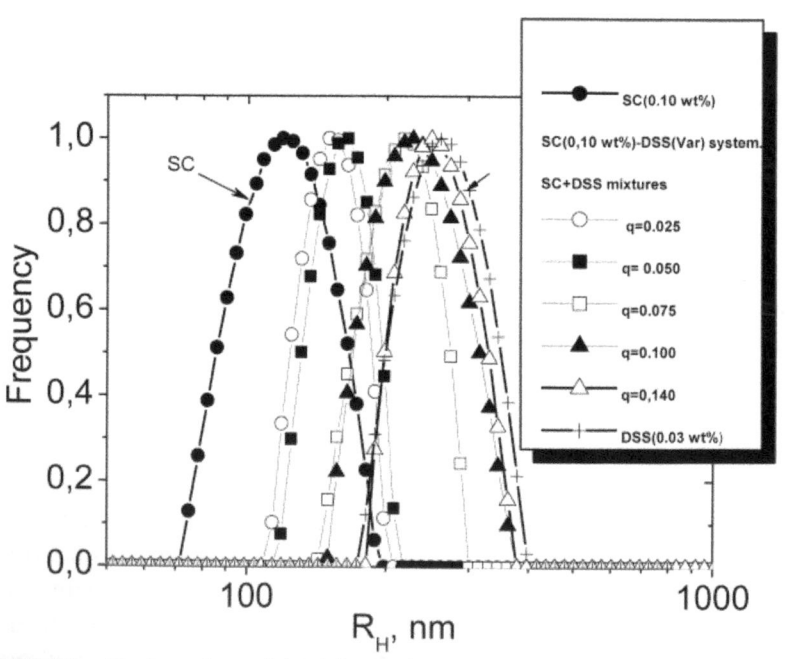

FIGURE 9.9 The intensity-weighted distribution of hydrodynamic radii (R_H) of solutions of sodium caseinate, dextran sulfate and their mixtures. Concentration of SC is equal to 0.1 (w/w). pH 7.0, $I = 0.002$ (phosphate buffer), 296 K.

An addition of DSS to SC solution at DSS/SC weight ratios ranging (q) from 0.025 to 0.05 leads to significant increase in the R_H toward the values R_H for DSS solution. At higher q values = 0.14, R_H of the mixed associates achieve the values R_H for DSS, and their size does not change with the further increase of q values. This is an indication of intermacromolecular interaction of the casein molecules with DSS and formation of complexes. At $q = 0.14$, function of the intensity-weighted distribution of hydrodynamic radii (R_H) is placed completely outside that describing free SC.

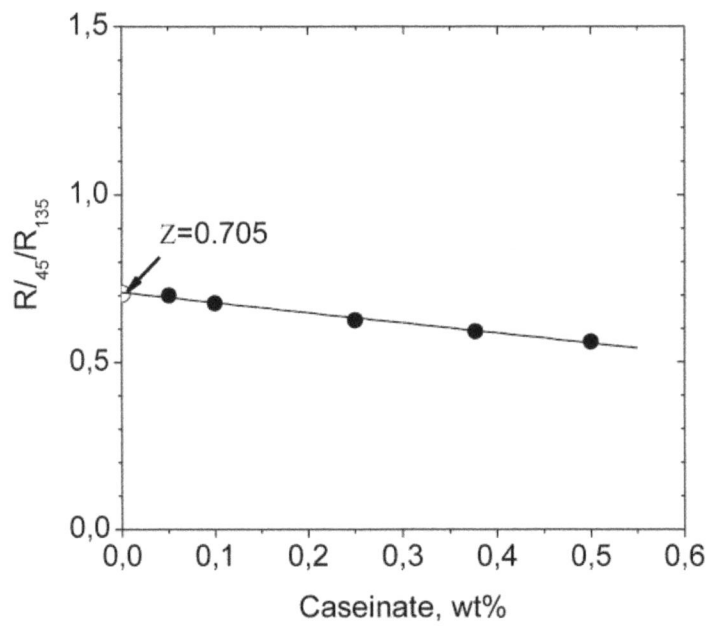

FIGURE 9.10 Dependence of the ratio of the scattering intensity, R at 45° and 135° on the concentration of casein in the SC–DSS mixture at $q = 0.14$. pH 7.0, $I = 0.002$ (phosphate buffer), 296 K.

The asymmetry coefficient (Z) of the complex associates was estimated by Debye method based on determination of the scattering intensity,(R) at two angles 45° and 135° symmetrical to the angle 90° and subsequent extrapolation of the R45°/R/135° to zero concentration. The results obtained are presented in Figure 9.10. The complex associates are asymmetric with Z values equal to 0.7.

Figure 9.11 presents zeta potential values and the total concentration of the biopolymer at the critical point C^t_{cr} as a function of the DSS/SC ratio, q. After an addition of DSS the negative value of the zeta potential increases and C^t_{cr} decreases achieving correspondingly the maximal and minimal values at $q = 0.14$.

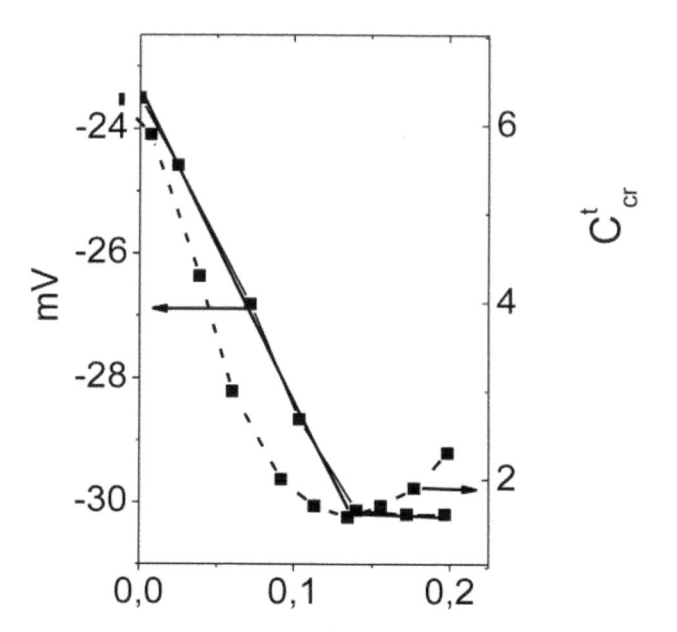

FIGURE 9.11 Dependence of zeta potential, and the total critical concentration of biopolymers corresponding to phase separation of W–SC–SA–DSS system on q. SC/SA weight ratio is 4. pH 7.0, $I = 0.002$ (phosphate buffer), 296 K.

Once the negative charge of a protein becomes higher in the presence of DSS, interactions between casein molecules could be hindered by an overall effect of electrostatic repulsion. Thus, an increase in the net charge of casein due to DSS binding could lead to an enhancement in the extent of such repulsions, contributing to the suppression of the further association and aggregation. Obviously, at pH = 7.0, the total charge of the high-molecular-weight DSS molecule is higher than the total positive charge of the relatively small SC molecule. This gives the possibility to regard complex formation between these biopolymers (similarly to other weak-polyelectrolyte—strong-polyelectrolyte interactions [54, 55]) as a mononuclear association in which the DSS molecule is the nucleus and the casein molecule is a ligand. Therefore, the formation of casein—DSS complex can be regarded as the reaction of few casein molecules successively joining one molecule of DSS nucleus. Note that at pH = 7.0, (experimental conditions) all the cationic groups of casein, as well as all the sulfate groups of DSS are ionized. It easy to show that at $q° = q*$ (0.135), the ratio of the total amount of sulfate groups in DSS molecule and cationic groups in casein molecule ($\frac{[S^-]}{[Cat^+]}$) is close to unity. Actually, the total amount of cationic groups in casein molecule is 0.76 mmol/g [50, 56] and the content of sulfur groups in DSS molecule is equal to 5.43 mmol/g [57]. Therefore, $\frac{[S^-]}{[Cat^+]} = q\frac{5.43}{0.76}$. At $q* = 0.135$, one can obtain $\frac{[S^-]}{[Cat^+]} = 0.964$.

Figure 9.12 presents the chromatograms of the initial solutions of SC (0.25 wt%) and DSS (0.25 wt%), and the SC–DSS system ($q = 0.14$, concentrations of SC = 0.25 wt% and 1.0 wt%), showing distribution of the protein, polysaccharide, and complex associates in the chromatographic fractions.

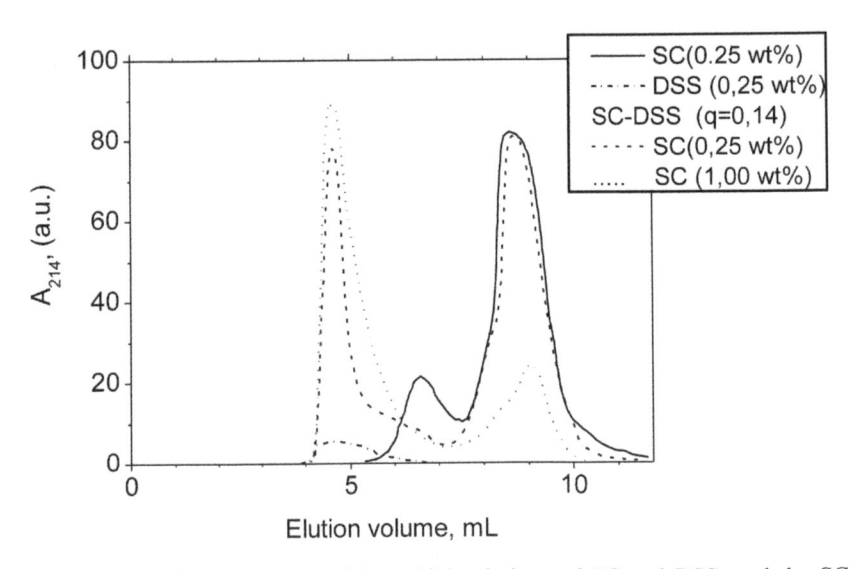

FIGURE 9.12 Chromatograms of the initial solutions of SC and DSS, and the SC–DSS system ($q = 0.14$), showing distribution of the protein, polysaccharide, and complex associates in the chromatographic fractions at concentrations of the protein 0.1 wt per cent and 1.0 wt per cent. pH 7.0, $I = 0.015$ (phosphate buffer), 296 K.

Free SC exhibited at pH 7.0 two unequal peaks. The first peak (83% from the total square) presents SC molecules, and the second one (17% from the total square) corresponds to the SC associates. Estimation of the molecular weights of these components on the basis of known molecular weights of alpha, beta, and gamma gelatins gave 260 kDa and 380 kDa accordingly. The weight average molecular weight of both fractions was about 300 kDa.

DSS exhibited at the same conditions a weak wide signal in the excluded volume. The chromatograms of the SC (0.25 wt%)-DSS systems at $q = 0.14$ gave a new high-molecular-weight component corresponding to excluded volume and the peak corresponding to the elution volume of the free (unbounded SC). It is interesting to note that at concentration of SC below the critical concentration of the phase separation, the degree of conversion of the SC in water soluble complex with DSS is low (30%), and mainly the high molecular fraction of SC interact with DSS. The interaction becomes stronger when the concentration of the SC in the mixture increases up to 1.0 wt% (inside two-phase range of SC–SA system in the presence of DSS ($q = 0.14$). In such conditions 83 per cent of SC form complex with DSS. Taking into account that the maximal yield of the complex takes place at $q = 0.14$, knowing the weight-average molecular weights of SC and DSS, and the degree of the protein conversion in protein–polysaccharide complex, we can roughly evaluate the SC/DSS molar ratio in the complex in the selected conditions corresponding to demixing of the mixed solutions of SC (2 wt%)-SA (0.5 wt%) in the presence of DSS ($q = 0.14$). Simple calculation showed that about 10 molecules of SC join to 1 DSS molecule, forming large associates with high molecular weight. Systematic experimental data concerning dependence of C^*_t upon the radius, or molecular weight of synthetic or natural polymers are unknown until now, although it is generally accepted that thermodynamic compatibility of polymers decreases with increase in molecular weights. It has been shown recently [58] that the total concentrations of biopolymers at the threshold point (C^*_t) for casein-guar gum system changes in accordance to $C^*_t \propto M^{cas}_w{}^{-0.27}$, where M^{cas}_w is molecular weight of caseins. This dependence has been established in a wide range of M^{cas}_w (from 25 kDa to 160.000 kDa). In that way, formation of large SC–DSS associates should decrease considerably compatibility of SA with bonded SC compared with that of "free" casein molecules that was observed in present work (Figure 9.4).

9.3.4 COMMONALITY OF THE DDS-INDUCED DEMIXING AT REST

The other question that arises from the demixing phenomenon in diluted biopolymer systems in the presence of DSS is, what is the key factor determining complex formation between DSS and caseins at pH 7.0 (far from the pH value corresponding to IEP of caseins)? Is the high local charge density of the positively charged kappa casein responsible for that, or it is mainly determined by the structural features of DSS, such as the concentration of sulfate groups, charge density, and conformation of the polysaccharide? Specific interaction between k-casein and carrageenan has been ascribed by [50]., to an attraction between the negatively charged sulfate groups of carrageenan and a positively charged region of κ-casein, located between residues 97 and 112.

It does not occur with the other casein types. Since the positive patch on κ-casein is believed to have a size of about 1.2 nm and is surrounded by predominantly negatively charged regions, the importance of the intersulfate distances is unmistakable. To extend, the Snoeren suggestion to our system, containing more stronger polyelectrolyte than carrageenan, or to reject it, we investigated the effect of DSS on the phase equilibrium in semi-dilute single-phase biopolymer systems containing the protein (gelatin) with the statistical distribution of the positively charged functional groups. Two systems were under consideration, gelatin type A–SA, and gelatin-type A–SC. The former is a single-phase one in water over a wide concentration range, and it undergoes phase separation at ionic strength above 0.2 [59].The latter system undergoes phase separation only at a very high ionic strength (above 0.5) [60] and is characterized by a very high total concentration of the biopolymer (>15–20 wt%) at the critical point [61].

The compatibility of these biopolymer pairs in water in the presence of DSS (at $q = 0.14$) was studies. The phase separation of both systems in the presence of DSS was established, and the binodal lines for them were determined (Figure 9.13). The binodals for the systems without DSS are placed outside the concentration range studied. In both systems the phase separation leads to formation of water in water emulsions with liquid coexisting phases (Figure 9.13). Two important conclusions can be made from these data. First, the DSS-induced phase separation in semi-dilute biopolymer solutions at rest is a rather general phenomenon not an exceptional case.

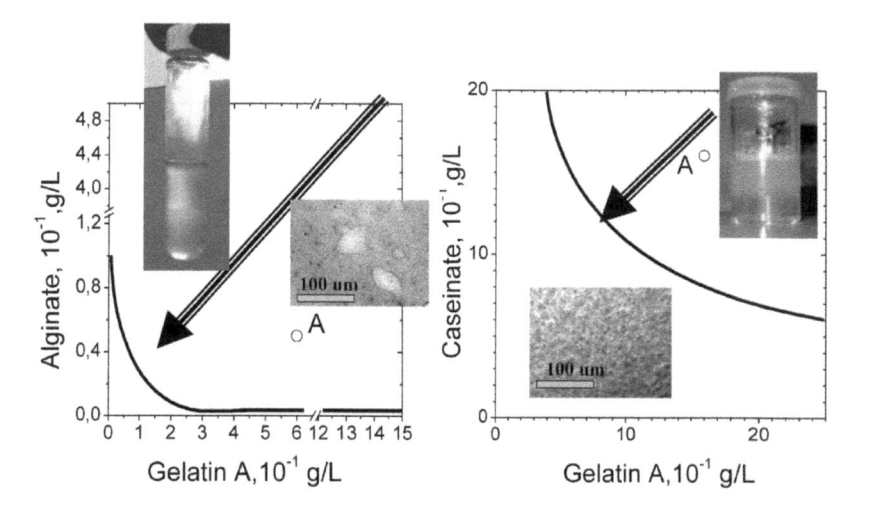

FIGURE 9.13 Shift of the Bimodal Line of the W-gelatin type A-SA, and W-SC-gelatin Type A Systems in the Presence of DSS at DSS/Protein Weight ratio 0.14; Photo Images and Microscopy Images of the Demixed W-gelatin type A(6 wt%)–SA (0.5 wt%), and W-SC (16 wt%)-gelatin type A (16 wt%) systems (points A on phase diagrams). W-gelatin type A-SA System was Prepared at pH 5.0, and W-SC-gelatin type A System was Prepared at pH 7.0.

Second, the structural features of DSS molecules is the more important factor determining complex formation of SC with DSS and subsequent demixing of the single-phase semi-dilute systems, rather than the characteristics of the distribution of the positively charged groups in the protein molecules. The last conclusion is in agreement with the FPLC data (Figure 9.12). As can be seen, the degree of the protein conversion in complex achieves 80 per cent, whereas the content of kappa casein in SA is only 12–14 per cent [62]. It is known that the sulfate groups of DSS are more closely packed than that of κ-carrageenan (0.5 nm for DSS and 1.2 nm for carrageenan [62, 63]. The latter can allow for the attractive forces to overcome the repulsive forces acting outside the positive patch. Bowman, Rubinstein, and Tan, characterizing complex formation between negatively charged polyelectrolytes and a net negatively charged gelatin by light scattering, suggested [64] that the protein is polarized in the presence of strong polyelectrolyte. Junhwan and Dobrynin have recently presented the results of molecular dynamics simulations of complexation between protein and polyelectrolyte chains in solution [65]. They found that protein placed near polyelectrolyte chains is polarized in such a way that the oppositely charged groups on the protein are close to the polyelectrolyte, maximizing effective electrostatic attraction between the two, whereas the similarly charged groups on the protein far away from the polyelectrolyte minimize effective electrostatic repulsion. In dilute and semi-dilute solutions, which are subjects of our study, polyampholyte chains usually form a complex at the end of polyelectrolyte chains resulting from the above polarization effect by polyelectrolyte. We believe that polarization-induced attraction is the main mechanism of complexation SC and DSS.

9.3.5 DISCUSSION ON THE STRUCTURE OF THE SC–DSS COMPLEXES AND SC ENRICHED PHASE OF THE DEMIXED SC–SA SYSTEM

From study of polyelectrolyte complexes, we know that interaction between oppositely charged polyelectrolyte's leads to partial or complete neutralization of charges, complexes remain soluble or precipitate, and in some cases gel-like networks are formed. If neutralization of charges is significant, the so-called "scrambled egg" compact structure will be formed. When neutralization of charges is far from complete, a "ladder" structure of complex can be formed [66].

The results of the zeta potential measurements, DLS, and flow experiments shown that the negative charge of the SC increases during interaction with DSS, and the maximal binding takes place at approx 0.14 DSS/SC weight ratio. Such features of the intermacromolecular interactions do not promote formation of the "scrambled egg" structure, because DSS molecule having many combined SC molecules and considerable negative charge can not be fold. Therefore, the ladder structure is more preferable for the system (Figure 9.14a). The overage size of the SC–DSS complex associates established from the DLS experiments is 0.2 um. Such a length scale would be in line with the fact that the SC/DSS solution is slightly turbid. This turbidity arises from a length scale in the micrometer range.

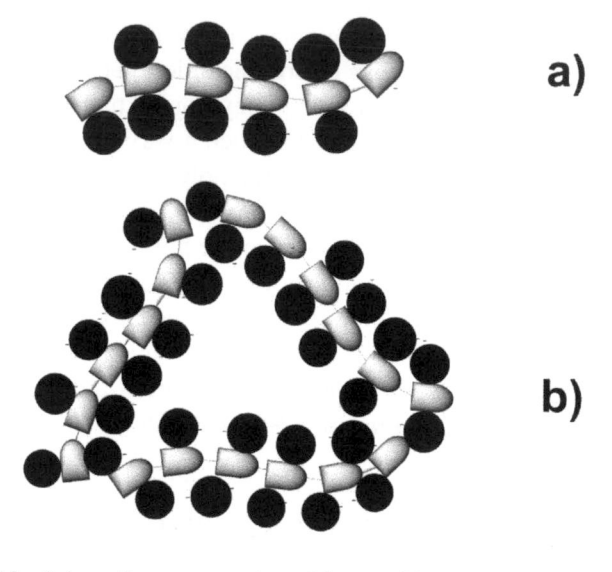

FIGURE 9.14 Schematic representation of the possible structures of (a) ladder-like and (b) gel-like. The long chain represent DSS molecule and the balls represent casein chains.

Obviously, heterogeneities on a micrometer scale were formed. If SC/DSS solution was made of a homogeneous structure of polymers on the nanometer scale, it would be transparent. In the presence of free polymer-SA, complex associates of SC and DSS undergo further association and the system becomes two phasic. This suggestion finds confirmation in the flow experiments; viscosity of the demixed SC–SA system is considerably higher than that of undemixed SC–SA system having the same concentrations (Figure 9.6). This difference is even much higher in the case of higher protein concentration in the single-phase SC–DSS system (Figure 9.8), this is a clear indication of association of the "ladder" structure of the complex associates, and formation of network (Figure 9.14b).

9.3.6 SHEAR-INDUCED BEHAVIOR OF THE SC–SA SYSTEM IN THE PRESENCE OF DSS

The experimental results shown in this section have been obtained on a water (97.5 wt%)-SC (2.0 wt%)-SA (0.5 wt%)-DSS ($2 \cdot 10^3$ wt%) system. It contains 99 wt per cent of the SC enriched phase and 1 wt per cent of the SA enriched phase, which have been mixed by hand, typically resulting in a very fine morphology. This emulsion is located in the two-phase region not far from the binodal line. The coexisting phases have Newtonian viscosities at 296 K, of 0.03 Pa·s and 0.02 Pa·s for the SC enriched and the SA-enriched phase, respectively.

To study the effect of flow on the phase behavior, a flow history consisting of three shear zones is used (Figure 9.15). First, a preshear of 0.5 s^{-1} is applied for 1,000 s (500 strain units). It has been verified that this procedure leads to a reproducible initial morphology. Subsequently, this preshear is stopped and the slightly deformed droplets are allowed to retract to a spherical shape. The resulting droplet radius is of the order of 5 micron. Finally, the shear rate is suddenly increased to a high value for 80 s, and after stopping flow the evolution of the SALS patterns are monitored.

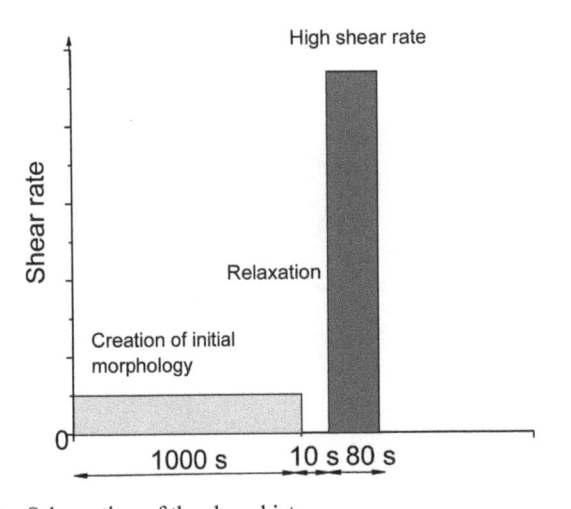

FIGURE 9.15   Schematic r of the shear history.

The evolution of the SALS patterns after cessation of steady-state shear flow at 60 s^{-1}, 100 s^{-1}, and 150 s^{-1} is shown in Figure 9.17. In each experiment, a freshly loaded sample has been used. As can be seen at all shear rates selected, the SALS patterns become more intensive just after cessation of flow. The higher the shear rate applied, the more intensive the SALS pattern becomes. This is a clear indication of shear induced demixing in SC–SA system in the presence of DSS. After cessation of shear flow, the light intensity is slowly decreasing (Figure 9.16), but the complete recovery of the initial SALS pattern takes place only after 1–2 h (data are not presented). In Figure 9.17 microscopy images corresponding to the same emulsion as in SALS experiments are presented first, after preshear of the emulsion at 0.5 s^{-1} for 1,000 s with subsequent cessation of steady-state shear flow at 60 s^{-1}(a), 100 s^{-1} (b), and 150 s^{-1}. One can see an appreciable increase of the droplet size after cessation of high shear rate flow, in accordance with SALS data.

In Figure 9.18, the light scattering intensity of semidilute demixed water (97.5 wt%)-SC (2.0 wt%)-SA (0.5 wt%)-DSS (2.10^{-3} wt%) system after preshear (curve 1) and just after cessation of flow at 60 s^{-1} (curve 2) is compared with that of water (87.8 wt%)-SC (12.2 wt%)-SA (0.1 wt%) system, containing 1 wt per cent SA enriched phase at the same shear history (curves 3 and 4). It is seen that the increase in the light intensity after cessation of flow takes place for both systems, however for the former system the light intensity increased much higher that for the latter one.

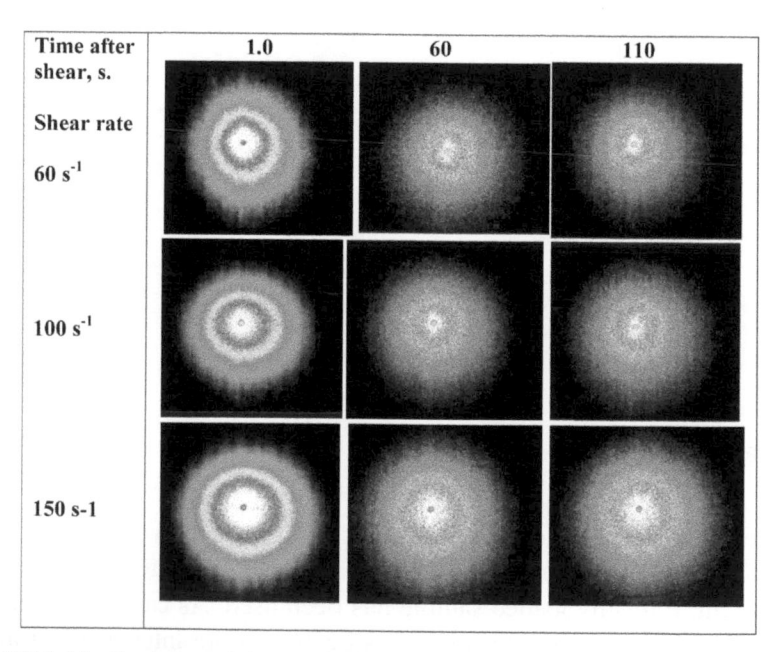

FIGURE 9.16 Evolution of the SALS patterns of water (97.5 wt%)-SC (2.00 wt%)-SA (0.5 wt%)-DSS (2 10^{-3} wt%) after cessation of a high shear rate flow. Shear rates and times of the shear as indicated on the figure. pH 7.0. $I = 0.002$ (phosphate buffer). Temperature 296 K. The SALS pattern of water (97.5 wt%)-SC (2.00 wt%)-SA (0.5 wt%)-DSS (2·10^{-3} wt%) system before high shear rate is shown in Figure 9.1a.

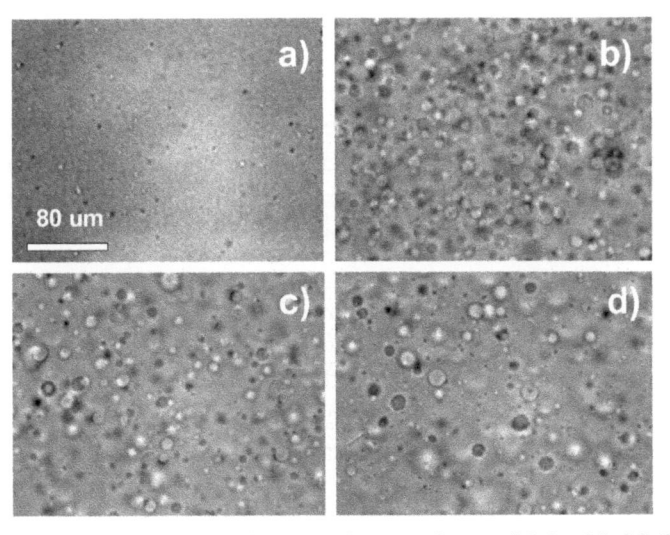

FIGURE 9.17 The evolution of microscopy images of water (97.5 wt%)-SC (2.00 wt%)-SA (0.5 wt%)-DSS (2 10^{-3} wt%) system before high shear rate flow (a) and just after cessation of a high shear rate flow. shear rate: b) 60 s^{-1}, c) 100 s^{-1}, and d) 150 s^{-1}. pH 7.0. I = 0.002 (phosphate buffer). Temperature 296 K.

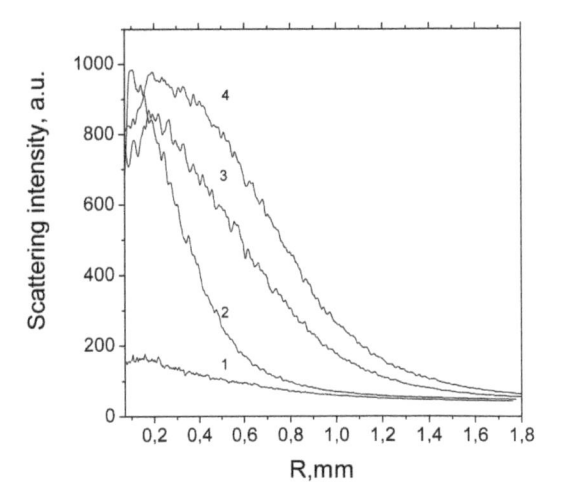

FIGURE 9.18 Dependence of the scattering intensity of the demixed water (97.5 wt%)-SC (2.00 wt%)-SA (0.5 wt%)-DSS (2·10^{-3} wt%) system after preshear (curve 1), and just after cessation of flow at 60 s^{-1} (curve 2) and water (87.8 wt%)-SC (12.2 wt%)-SA (0.1 wt%) system after preshear (curves 3), and just after cessation of flow at 60 s^{-1}, on the distance from the bean stop. Both systems contain 1.0 wt per cent SA-enriched dispersed phase.

These observations can be explained on the basis of a comparison of the molecular weights of the "free" SC and SC, combined with DSS (see Figure 9.12). The molecular weight of the latter one is much higher than that of the former one. Note that the second virial coefficients on the molar scale, related to pair interactions of similar SC macromolecules, A_{22} depends on the molecular weight inversely [67]. Therefore, according to conditions of the phase separation in biopolymer systems in flow [68]:

$$A_3 > \sqrt{A_2 \, A_3} \qquad\qquad (9.2)$$

in which A_{ij} are the second virial coefficients on the molar scale, related to pair interactions of similar (2-protein, 3-polysaccharide) and dissimilar macromolecules, the protein-polysaccharide mixture containing macromolecules with lower values of A_{22} will be more predisposed to shear induced demixing.

9.4 CONCLUSIONS

It well known that phase equilibrium in aqueous system containing casein and linear acid polysaccharide is weakly sensitive to changes of the main physico-chemical parameters, such as pH, ionic strength, and temperature. This is the case both at rest [17, 31, 32, 34] and under shear flow [69]. The weak intermacromolecular interactions caused by the presence of a complexing agent in two phase biopolymer mixture can affect its phase equilibrium and morphology. In this communication, the attempt was performed to induce demixing in semidilute and highly compatible sodium caseinate/sodium alginate system (SC–SA) mixtures in the presence of sodium salt of dextran sulfate (DSS) at pH 7.0, (above the isoelectrical point of caseins), and to characterize phase equilibrium, intermacromolecular interactions, and structure of such systems by rheo-small angle light scattering (SALS), optical microscopy (OM), phase analysis, dynamic light scattering (DLS), fast protein liquid chromatography (FPLC), ESEM, and rheology.

Addition of dextran sulfate sodium salt (DSS) to the semi-dilute single phase SC–SA system, even in trace concentrations (10^{-3} wt%), leads to segregative liquid–liquid phase separation, and a substantial increase in storage and loss moduli of the system. The degree of the protein conversion in the complex grows, when the concentration of SC in the system increases from 1 to 2 wt per cent. It is also established here that demixing of semi-dilute biopolymer mixtures, induced by the minor presence of DSS is a rather common phenomenon, because it is also was observed here for other biopolymer pairs. At high shear rates SC becomes even less compatible with SA in the presence of DSS than at rest. Experimental observations suggest that the approach for inducing demixing of semidilute and highly compatible biopolymer mixtures by physical interactions of the constituents is a promising tool for regulation of biopolymer compatibility and achieving better predictions of phase behavior of aqueous protein-charged polysaccharide systems.

KEYWORDS

- **Biopolymer mixture**
- **Complex formation**
- **Demixing**
- **Rheo-optics**
- **Structure formation**

REFERENCES

1. Hidalgo, J.; Hansen, P. M. T. J.; *Dairy Sci.* **1971**, *54*, 1270–1274.
2. Wang, Y.; Gao, J. Y.; Dubin P. L.; *Biotechnol. Prog.* **1996**, *12*, 356–362.
3. Dubin, P. L.; Gao, J.; Mattison, K.; *Sep. Purif. Methods*, **1994**, *23*, 1–16.
4. Strege, M. A.; Dubin, P. L.; West, J. S.; Flinta, C. D.; Protein separation via polyelectrolyte complexation. In *Symposium on Protein Purification: From Molecular Mechanisms to Large-scale Processes*. Ed. Ladisch, M.; Willson, R. C.; Paint C. C.; Builder, S. E.; ACS Symposium Series 427, Chapter 5. Washington: American Chemical Society; **1990**, 66 p.
5. Antonov, Yu. A.; Grinberg, V. Ya.; Zhuravskaya, N. A.; Tolstoguzov, V. B.; *Carbohydrate polymers*, **1982**, *2*, 81–90.
6. Antonov, Yu. A.; Application of the Membraneless osmosis method for protein concentration from molecular-dispersed and colloidal dispersed solutions. *Review. Applied Biochem. Microbiol.* (Russia) Engl. Transl., **2000**, *36*, 382–396.
7. Kiknadze, E. V.; Antonov, Y. A.; *Appl. Biochem. Microbiol.* (Russia) Engl. Transl., **1998**, *34*, 462–465.

8. Xia, J.; Dubin P. L.; Protein-polyelectrolyte complexes In: Macromolecular Complexes in Chemistry and Biology, Ed. Dubin P. L.; Davis R. M.;. Schultz, D.; Thies C.; Berlin: Springer-Verlag, **1994**, Chapter 15.

9. Ottenbrite, R. M.; Kaplan, A. M.; *Ann. N. Y. Acad. Sci.* **1985**, *446*, 160–168.

10. Magdassi, S.; Vinetsky, Y.; Microencapsulation: methods and industrial applications. In Microencapsulation of oil-in-water Emulsions by Proteins. Ed. Benita, S.; New York: Marcel Dekker Inc.; **1997**, 21–33.

11. Regelson, W.; *J Bioact. Compat. Polym.* **1991**, *6*, 178–216.

12. Albertsson, P.-Å.; Johansson, G.; Tjerneld, F.; In Separation Processes in Biotechnology, Ed. Asenjo, J. A.; New York: Marcel Dekker, **1990**, 287–327.

13. Harding, S.; Hill, S. E.; Mitchell, J. R.; Biopolymer Mixtures. Nottingham: University Press; **1995**, 499 p.

14. Tolstoguzov, V. B.; Functional properties of protein-polysaccharide mixtures. In: Functional Properties of Food Macromolecules. Ed. Hill, S. E.; Ledward, D. A.; Mitchel, J. R.; Gaitherburg, Maryland: Aspen Publishers Inc.; **1998**.

15. Piculell, L.; Bergfeld, K.; Nilsson, S.; Factors determining phase behaviour of multicomponent polymer systems. In Biopolymer Mixtures. Ed. Harding, S. E.; Hill, S. E.; Mitchell, J. R.; Nottingham: Nottingham University Press; **1995**, 13–36.

16. Antonov, Y. A.; Van Puyvelde P.; Moldenaers, P.; *Biomacromolecules*, **2004**, *5(2)*, 276–283.

17. Antonov, Y. A.; Van Puyvelde P.; Moldenaers, P.; *IJBM*, **2004**, *34*, 29–35.

18. Antonov, Y. A.; Wolf, B.; *Biomacromolecules*, **2006**, *7(5)*, 1582–1567.

19. Walter, H.; Brooks, D. E.; Fisher, D. In Partitioning in Aqueous Two-Phase Systems: Theory, Methods, Uses and Applications to Biotechnology. London: Academic Press; **1985**.

20. Hellebust, S.; Nilsson, S.; Blokhus, A. M. *Macromolecules*, **2003**, *36*, 5372–5382.

21. Scott, R. J.; *Chem. Phys.* **1949**, *17*, *3*, 268–279.

22. Tompa, H.; *Polym. Solut.* London: Butterworth; **1956**.

23. Flory, P. J.; Principles of Polymer Chemistry. Ithaca, NY: Cornell University Press; **1953**.

24. Prigogine, I.; The Molecular Theory of Solution. Amsterdam: North-Holland; **1967**.

25. Patterson, D.; *Polym. Eng. Sci.* **1982**, *22* (2), 64–73.

26. Gottschalk, M.; Linse, P.; Piculell, L.; *Macromolecules.* **1998**, *31*, 8407–8416.

27. Lindvig, T.; Michelsen, M. L.; Kontogeorgis, G. M.; *Fluid Phase Equilib.* **2002**, *203*, 247–260.

28. Piculell, L.; Nilsson, S.; Falck, L.; Tjerneld, F.; *Polym. Commun.* **1991**, *32*, 158–160.

29. Bergfeldt, K.; Piculell, L.; Tjerneld, F.; *Macromolecules.* **1995**, *28*, 3360–2270.

30. Bergfeldt, K.; Piculell, L. J.; *Phys. Chem.* **1996**, *14*, 5935–5840.

31. Antonov, Y. A.; Grinberg, V. Ya.; Tolstoguzov, V. B.; *Starke.* **1975**, *27*, 424–431.

32. Antonov, Y. A.; Grinberg, V. Ya.; Zhuravskaya, N. A.; Tolstoguzov, V. B. J.; *Texture Studies.* **1980**, *11(3)*, 199–215.

33. Rha, C. K.; Pradipasena, P.; In Functional Properties of Food Macromolecules. Ed. Mitchell, J. R.; Ledward, D. A.; London: Elsevier Applied Science; **1986**, 79–119.

34. Whistler, R. L.; Industrial Gums, 2nd edn. New York: Academic Press; **1973**.

35. Capron, I.; Costeux, S.; Djabourov, M.; *Rheol. Acta.* **2001**, *40*, 441–456.

36. Van Puyvelde, P.; Antonov, Y. A.; Moldenaers, P.; *Food Hydrocolloids.* **2003**, *17*, 327–332.

37. Koningsveld, R.; Staverman, A. J.; *J. Polym. Sci. A-2,* **1968**, 305–323.

38. Polyakov, V. I.; Grinberg, V. Ya.; Tolstoguzov, V. B.; *Polym. Bull.* **1980**, *2*, 760–767.

39. Dubois, M.; Gilles, K. A.; Hamilton, J. K.; Revers, P. P.; Smith, T.; *Anal. Chem.* **1956**, *18*, 350–356.

40. Zaslavsky, B. Y.; Aqueous Two Phase Partitioning: Physical Chemistry and Bioanalytical Applications. New York: Marcel Dekker; **1995**.

41. Grinberg, V. Ya.; Tolstoguzov, V. B.; *Food Hydrocolloids.* **1997**, *11*, 145–158.

42. Van Puyvelde, P.; Antonov, Y. A.; Moldenaers, P.; *Food Hydrocolloids.* **2003**, *17*, 327–332.

43. Overbeek, J. T. G.; Voorn, M. J. J.; *Cell Comp. Physiol.* **1957**, *49*, 7–39.

44. Kaibara, K.; Okazaki, T.; Bohidar, H.P.; Dubin, P.; *Biomacromolecules.* **2000**, *1*, 100–107.

45. Antonov, Yu. A.; Dmitrochenko, A. P.; Leontiev, A. L.; *Int. J. Biol. Macromol.* **2006**, *38 (1)*, 18–24.

46. Antonov, Y. A.; Gonçalves, M. P.; *Food Hydrocolloids.* **1999**, *13*, 517–524.

47. Michon,C.; Konate, K.; Cuvelier, G.; Launay, B.; *Food Hydrocolloids.* **2002**, *16*, 613–618.

48. Galazka, V. B.; Ledward, D. A.; Sumner, I. G.; Dickinson, E.; *Agric. Food Chem.* **1997**, *45*, 3465–3471.

49. Dickinson, E.; *Trends Food Sci. Technol.* **1988**, *9*, 347–354.

50. Snoeren, T. H. M.; Payens, T. A. J.; Jevnink, J.; Both, P. *Milchwissenschaft.* **1975**, *30*, 393–395.

51. Kabanov, V. A.; Evdakov, V. P.; Mustafaev, M. I.; Antipina, A. D.; *Mol. Biol.* **1977**, *11*, 582–597.

52. Noguchi, H.; *Biochim. Biophys. Acta.* **1956**, *22*, 459–462.

53. Kumihiko, G.; Noguchi, H.; *Agricul. Food Sci.* **1978**, *26*, 1409–1414.

54. Sato, H.; Nakajima, A.; *Colloid Polym. Sci.* **1974**, *252*, 294–297.

55. Sato, H.; Maeda, M.; & Nakajima, A.; *J. Appl. Polym. Sci.* **1979**, *23*, 1759–1767.

56. Bungenberg de Jong, H. G.; Crystallisation-Coacervation-Flocculation. In Colloid Science. Ed. Kruyt, H. R.; : Amsterdam: Elsevier Publishing Company; **1949**, Vol. II, Chapter VIII, 232–258.

57. Gurov, A. N.; Wajnerman, E. S.; Tolstoguzov, V. B.; *Stärke.* **1977**, *29* (6), 186–190.

58. Antonov, Y. A.;Lefebvre, J.; Doublier, J. L. *Polym. Bull.* **2007**, *58*, 723–730.

59. Antonov, Y. A.;Lashko N. P.;.Glotova, Y. K.; Malovikova, A.; Markovich, O. *Food Hydrocolloids.* **1996**, *10 (l)*, 1–9.

60. Polyalov,V. I.; Grinberg, V. Ya.; Antonov, Y. A.; Tolstoguzov, V. B.; *Polym. Bull.* **1979**, *1*, 593–597.

61. Polyalov, V. I.; Kireeva, O. K.; Grinberg, V. Ya.; Tolstoguzov, V. B.; *Nahrung.* **1985**, *29*, 153–160.

62. Swaisgood, H. E.; In Advanced Dairy Chemistry-1:Proteins. Ed. Fox, P. F.; London: Elsevier Applied Science; **1992**.

63. Langendorff, V.; Cuvelier, G.; Launay, B.; Michon, C.; Parker, A.; De Kruif, C. G.; *Food Hydrocolloids.* **1999**, *13*, 211–218.

64. Bowman, W. A.; Rubinstein, M.; Tan, J. S.; *Macromolecules.* **1997**, *30(11)*, 3262–3270
65. Junhwan, J.; Dobrynin, A.; *Macromolecules.* **2005**, *38(12)*, 5300–5312.
66. Smid, J.; Fish, D.; In: Encyclopedia of Polymer Science and Engineering, 2nd ed. Polyelectrolyte Complexes. Ed. Mark, H. F.; Bikales, N. M.; Overberger, C. G.; Menges, G.; New York: Wiley/Interscience; **1988**, Vol. 11, 720 p.
67. Striolo. A.; Ward, J.; Prausnitz, J. M.; Parak, W. J.; Zanchet, D.; Gerion, D.; Milliron, D.; Alivisatos, A. P. J.; *Phys. Chem.* **2002**. B, *106(21)*, 5500–5505.
68. Antonov, Y. A.; Van Puyvelde P.; Moldenaers, P.; *Biomacromolecules.* **2004**, *5(2)*, 276–283.
69. Antonov, Y. A.; Van Puyvelde, P.; Moldenaers, P.; *Food Hydrocolloids.* **2009**, *23*, 262–270.

INDEX